Web3 and Fintech in Business

What you need to know about Blockchain, DeFi, Cryptocurrency and IoT in the current Business Environment

By

Dr. Christopher W. Smithmyer

Elite exclusivity

Dysart Detroit Dallas

First Edition 2024

Send All Inquiries to: Info@EliteExclusivity.com

ISBN Paperback: 978-1-952651-40-3

ISBN Hardback: Forthcoming

ISBN Digital: Forthcoming

Printed in the United States of America

Contents

Introduction

The financial world is changing. While this is an obvious statement, because the financial world is always changing, when you are dealing with the fintech world- you need to know that this change is not just inevitable, it is also part of the fabric of the system. Fintech thrives on the fact that new methodologies and information is being created moment to moment. Where financial systems in the TradFi (the term for traditional finance) may last for centuries, some concepts within the fintech system exist only for days before they are improved upon and outdated. This is a world of adventure and exploration.

This book is designed to be offered in conjunction with the basics of cryptocurrency, blockchain and fintech course which is taught by Dr. Christopher W. Smithmyer, Abdullah Jadoon, and others. While the information in the book is stand alone, the book acts as a guidebook- a map of sorts- as the reader engages upon the blockchain to explore the spiderweb bridge which is the cryptoverse. Our goal, is to engage the reader with the facts and the nature of the cryptoverse while attempting to keep this from being a chore to read. Anytime one is writing about economics, this can be a challenge.

> ***Cryptoverse****: The collection of blockchains, cryptocurrency, the Internet of Things (IoT) fintech services and digital financial systems which operate within the blockchain.*

To this end, we have broken the book into short, readable chapters that an active person can read on the go, or a person who has the time to sit down and read the material all at once can enjoy without the concepts being too broken up. The goal, therefore, is an economics Dan

Brown style book, which works with you on your schedule- and does the same for your neighbor. Additionally, we are going to attempt to introduce new terminology slowly, through text boxes, like the one to the right, so that you can learn as you go what different words within the cryptoverse mean.

Methodology

This book is written as an executive summary of issues in the cryptoverse. To that end, the goal of this book is to keep each chapter between two or three pages to ensure that the reader has the ability to access the information that is needed on a given subject, so they can converse intelligently, then to allow them to do their own research on the subjects in which they are most interested. The chapters are designed to be able to be read very quickly, in the minutes before a meeting, on the bus, or when you just have a couple of minutes and you want to expand your knowledge. Other books in this group (Dragons of the Digital Age, Digital Exchequer, the Stablecoin Book, the Tome of Tokenomics, etc.) provide with more in-depth analysis of specific topics; whereas, this book is designed to be a survey of the subject for the reader.

Additionally, this book was AI edited. The authors wrote their sections, some coming in at much more than 3 pages, and then asked the AI to edit the chapter down to the proper word count. This has compressed the data to a level where it can work within the handbook fashion. As a result, some of the text will be very dense as to the subject matter. To alleviate this, we have attempted to have the above mentioned definitions in the sidebars and have an index and glossary at the end of the book to enhance the readability. Once the chapter has been reduced to the proper number of words, the authors re-examine the chapter and add the citations as needed (which are stored from the original text). This allows the text to be a human/AI hybrid which relies on the writing ability of the author and the editing ability of the author when they go over the AI reduced brief.

Parts of the Book

Part I of this text is a preparatory portion of the book, the chapters therein. This is an overview of the basics, striving to help those that would undertake the journey of moving their business, at least partially into the cryptoverse. Understanding the philosophy of the cryptoverse is an important first step in building up a plan for cryptocurrency penetration, like the culture of a nation to which you are attempting to incorporate into your business family is important for business development, so too is the culture of the cryptoverse as you attempt to build business enclaves in the digital frontier. This continues into a discussion of "what is blockchain? Which helps us look at the framework of the cryptoverse through the main thoroughfares which run through it. From there, the topic turns to the tools that you will need to embark on this journey, this is a working definition of what cryptocurrency is and how it can be used. In our case we need a deeper understanding of the difference between coins and tokens. This basic element of the cryptoverse allows us to see the different levels of utility which the cryptoverse provides.

Part I continues with a conversation about the on-ramps and off-ramps of the cryptoverse. This is the methodology professionals can use to enter and exit the cryptoverse. From there we look at systems that are permissionless and zero-trust, two of the key elements that allow the cryptoverse to exist as an immutable force within the business community. We then turn our attention to the concept of decentralization, which is the process of taking control away for a single entity or small group and placing the governance across the community. This is another core value of the cryptoverse, empowering millions across the world to control their own wealth. The next engagement is the wild west of the fintech regulatory structure, and the slow progress of civilization into the area. As business moves in, we see less speculation and more investment in the space. Finally, Part I concludes with a holistic view of the cryptoverse and the business opportunities on the horizon.

Part II is an analysis of the blockchain network, an series of nodes in the cryptoverse connected by an amazing spiderweb bridge of blockchains, exchanges, and digital businesses. Part II begins with a look at smart contracts and their importance to the cryptoverse. While the progenitor of cryptocurrency, Bitcoin, does not have direct smart contract functionality, the altcoins bring different levels of utility to the cryptoverse with the advent of smart contract protocols. Attention then shifts to decentralized Finance (DeFi), which is the moving of finance away from a core system and towards a system which is distributed across the market. One of the core elements of this is tokenization, digitization and valuating of goods and services so that they can be accurately represented on the blockchain. The conversation then shifts to supply chain integration and identity management. Because tokenized assets are stored on the blockchain, the access to said resources needs to be carefully managed. Likewise, the identity of the users must be continually confirmed to ensure that those who are to be using the resources are the only ones with access to said resources.

Part II continues with a brief look at cross border payments, while this topic is covered in detail later in the book, cross border payments function as one of the key areas of interest in cryptocurrency for business. In order for business, especially cross border payments, to be successful, blockchain needs to be set up that there is a high level of data security and privacy within the system. This integration is made all the more important when looking at the connectivity of the Web3 system, such as the Internet of Things (IoT). This is the network of devices and equipment which we use in the world which are constantly connected to and working with the Web3 universe. Part II then dips its toes into the realm of Non-Fungible Tokens, looking at how the gaming world has shown utility in this phenomenon well beyond the sharing of artistic creations. Part II concludes with a review of RegTech, a survey of the regulatory compliance issues within the Web3 space.

Part III looks at one of the key areas which interests most business in the Web3 space, how blockchain can be used to allow foreign market penetration. To start off, the discussion must look at the present (traditional) system of cross-border payments and how they function, along with some of their shortcomings. From there, the discussion turns to the concept of how Web3 and cryptocurrency facilitate international trade, moving beyond such simple systems as SWIFT. Visibility then becomes a key issue in the development of Web 3 in the international space, for without transparency there is a question as to the validity of the system. Initially, this was developed through self-actuating contracts (smart contracts); however, this has developed into NFTs which are designed to memorialize transactions and provide receipts.

One of the greatest difficulties of the Web3 space, especially on the international level is complying with the regulations. The TradFi market structure developed over the course of thousands of years, the Blockchain world is on 15 years old (30 at the most). This means that there are many new areas which still need to be explored. In this conversation, we then return to the concept of tokenization of assets. At the higher level, international transactions can be looked at as assets, thus tokenized for ease of use. With tokenization comes comprehension, which means that Web3 must address language and cultural issues within the space; however, the binary language of the computers does not care about these things-but there is a need for the system to take it into account for the human interface. Building international teams and reaching global audiences run hand in hand to complete Part III. International teams allow for better access to regulatory understanding and provide a more decentralized workforce. In order to do this, and build an international customer base, it is imperative to have strong global outreach.

Part IV begins with a primer to web3 as terminology within the world of business. While Web3 creates features which revolve around many of the aspects of modern life, the primary focus in this area will be business relationships. One specific relationship which cannot be understated is how Web3 and blockchain are interconnected in the development of humanities digital frontier. This is augmented by the advancement of Artificial Intelligence, which has made a paradigm shift in business and computing. The chapter goes on to look at how the roadmap for DeFi has shown that many other elements of individual life can be decentralized, made more efficient and for a lower cost. This includes, but is not limited to embedded finance, where purchases and savings are embedded into every aspect of life.

In the second half of Part IV, we look into the discussion of open banking, what it means, especially what it means to business and how it can affect every day decisions. Hand in hand with this is the topic of insuretech, which is how we can digitize insurance to ensure that people are covered for the troubles they face in life. This leads to a reprise of regtech, specifically how regtech needs to be constantly evolving with the system to ensure that it is not hampering the system. Taking a step back from Web3 (kind of), we must also look at how we can humanize the transactions within the system, avoiding the prospect of becoming too digital and facing a WALL-E crisis. Part IV concludes by looking at digital wallets, their types, and how our Web3 lives are stored in these interesting devices.

Part V looks at the policies of cross border payments. This part of the book expands on the topics covered in Chapter 16. Starting with the basic concept of cross border payments, we quickly move on to discuss past iterations of global payment systems. These include SWIFT, SWIFT 2.0, SWIFT GPI and Ripple/XRP. In each of these case, we look at the strengths and failures of the system and how they function within the global exchange rubric.

Moving beyond past systems, Part V then turns its attention to the general utility of modern cryptocurrencies, almost all of which self-validate. The topic then shifts to how this deals with jurisdictional problems in cross-border payments, troubles centralized systems such as SWIFT or XRP are still dealing with that are avoided by decentralized systems. As such, the development should move from a regulatory system where decentralization allows one to operate- to a system which is designed to support the decentralized system of finance.

Functionality and security are very important, chief among these areas is settlement functionality. The speed and validity of a transfer across borders is paramount as one looks at what system to use. Included in this functionality is the necessity for traceability, business requires that transfers can be tracked and validated- especially when trying to prevent their business from being used by bad actors. Finally, Part V looks at the delicate dance of privacy and KYC as we attempt to stabilize the world and ensure that problems are not enhanced by Web3 and cryptocurrency (such as they were with Web2 and the Dark Web).

Part VI delves into challenges, some of them alluded to in previous parts. Regulations are always going to be a balancing act when dealing with cryptocurrency, which was designed to decentralize finance, something regulations undercut. However, regulations are needed to ensure that there are basics security conditions which will allow for the system to be trusted. Intrinsically, this means that the field needs to acquire, both externally and internally, the best talent in IT, business, and law to ensure that new systems meet the requirements of the law, the vibrancy of business, and the technicality of code. As an overarching element, the technological infrastructure globally needs to be improved to allow access to the underserved and the underbanked.

Moving from general issues to specific business issues, one of the chief problems facing the community is the understanding of integration costs. Like the advent of electricity, the movement to Web3 requires updating of systems, new technology, and staff capable of handling that technological change. This includes taking the legacy systems which are currently in place and doing a practical audit of the system to determine whether the system is functional, usable, or something that should be scrapped for the new systems.

Auditing technology within the confines of a company is one thing; however, there also needs to be a customer base audit as well. Customers have been programmed since the creation of the radio to expect certain things out of business. While some of these things have been able to change over the course of time, it costs billions of dollars to 'change the programing.' Additionally, there needs to be an analysis of how your market competition is changing to meet the needs of the system- are they evolving with you or are they staying in the old world hoping to control the stragglers. Industrywide, this can be a problem of change resistance, something that is normal in the field. Finally, after all of these considerations are taking into account, any business adopting Web3 technology needs to look at a scalability plan and how it will function within the overarching system.

Part VII looks at the tools which we need as a system to build a better cryptoverse. Much like the boomtowns of the wild west, the cryptoverse is yearning for order and civilization. Clear and flexible regulations are one of the first elements that are needed to ensure civilization begins to move into the cryptoverse. To do this, we need regulatory sandboxes which allow for controlled environment to be established for regulatory experimentation. These sandboxes should call for cooperation between regulators and businesses, letting the decision makes talk directly to those who will be the benefits of their decisions.

Any region of the world which hopes to be a leader in the space needs to ensure that there is a strong program for fintech and cryptocurrency education. People are more hesitant to participate in something that they do not understand. As a result, they need to have adult classes on the subjects along with having the subjects integrated into the high school (grad school) and university curriculum. Countries also need to make an effort to attract outside talent. Unfortunately, the coming digital exploration will be both collaborative and competitive, this means that the best minds will be able to 'name their price' as the field develops. One way to do this is to ensure regional countries build a strong network of technology that transcends borders to allow those who come to the region to work to have access to top of the line equipment.

When looking at development at the national level, tax incentives and financial support are paramount for attracting the best business to a country (or city). Companies are looking for areas where they will have clear regulatory systems and a financial incentive to come and stay. Another methodology which has been shown to be successful is networking and collaboration hubs, incubators, and startup launch organizations. This is ultimately enhanced by open data an accessibility by all, which increase the number of actors in the space. The ultimate goal of any program to embrace Web3 is to enhance the standard of living for everyone in the nation.

Part VIII is an internal playbook for developing your ability as a company or a nation to explore the cryptoverse. One of the first goals of any business or country in this space should be to create a fintech team or working group. This group should be people in leadership positions who can look at specific fintech areas to address and suggest them to the group. The most basic effort for a Web3 novitiate company or nation is to build an inhouse Web3 system, while in most cases this system will be temporary, it allows the team to work hands on with Web3 data and build the experience needed to make the system successful. Once you have the knowledge to do so, it is useful to integrate fintech at some level into all departments, even if it is a third party accounting system. This allows the entire team to get used to blockchain systems.

Once a company is comfortable with fintech and blockchain in operations. Then the fintech team should be expanded to a fintech/Web3 research and development division. Following in the Web3 space is profitable, leading, however, can be a game changer. This includes brining in outside groups to do employee training sessions, creating an internal adoption roadmap for internal and external fintech systems, incentivizing innovation among your employees and implementing cybersecurity measures. Collaboration with other companies that are in the space is also a useful tool for development, allow different actors in the group to learn from others mistakes so that their mistakes are new and others can learn from them. Finally, create a commission with your peers for a fintech solution group specific to your industry. This will help move you beyond the basics to something that works for your individual industry or a market segment within a national industry.

Part IX deals with the fear that is out there revolving around the cryptocurrency world. Banks, nations, and wealthy individuals who do not see the benefit of cryptocurrencies are spreading fear of missing out (FOMO) about the industry. This can cause people to have concerns, rightfully. Addressing the concerns starts with an education and awareness campaign for regular people- even if they are not involved with the cryptocurrency industry. Communities should have regulatory clarity and governments should have support systems for people who are wronged by bad actors.

Privacy concerns are one of the 'boogie-men' of the cryptocurrency world. People have a misunderstanding that governments and the internet do not have access to what happens on the internet. Truth be told, people need to learn that major websites are farming customers for data- which in many cases makes more money for the company than sales. Once this is understood, the immutable ledger of the cryptoverse seems less scary. People also need to be introduced to the idea that fintech companies are promoting financial inclusion, they are not exclusionary- each new customer is a new account where profit can be generated. To help with this understanding, there needs to be a user-friendly interface that mitigates security risks. People expect the company to balance security and ease of access so they can use the website/app. The more we see TradFi institution and local governments integrate Web3 and cryptocurrency into their system, the more we will see people adopt the system. Consumers, for the most part, are smarter than people give them credit for- thus transparency and trust will help create more customers than fly-by-night programs.

Part X is the shortest section of the book and deals with advanced cryptocurrency elements. First and foremost, one needs to realize that the cryptoverse is a system. Though there are many pathways through the system (blockchains), the spiderweb bridge of the system connects everything. This means that the markets of the system will always affect cryptocurrency in one way or another- even if we do not see the relationship. Within the cryptoverse, we see the hubs of activity built around retail usage- most popular among these are centralized exchanges and decentralized exchanges (DEX). These are the two primary vehicles that act as connections between different chains (there are other portal) and create a visible level of connectedness even as we seek to understand the connections on the quantum level.

Part XI looks at NFTs. Understanding the utility of NFTs means looking at the underlying purpose of the system. To do this, we need to look at the utility of the system as a digital receipt, the title for items, or as memorials. We have moved beyond the basic elements of NFT design, where they were mere representations of art, to a point where they can be used as independent smart contracts and have self-executing functionalities.

SFTs bridge the gap between traditional tokens and NFTs. Part XII explains the methodology of the SFT system before engaging in a use case scenario analysis. SFTs are useful because of their dual nature, both interchangeable and specific- depending on the viewpoint of the users. SFTs excel in the area of stock development, perhaps only overshadowed by their ability to represent derivative ownership. Like NFTs they bring the ability for tokenized assets to be traced and verified across the system; however, like traditional tokens they can be openly traded without a care for which specific SFT someone is getting from a specific pool.

We hope therefore that you enjoy this engagement into the realm of Web3 and cryptocurrency. If you are studying with us at one of our global classes, the more your read before the class the more useful the class will be for you. However, you can still engage in the class without reading the book, or engage with the book without taking the class. Your exploration of the cryptoverse is on your terms, hopefully this will be a good guidebook on your journey.

Part I: Navigating the Seas of Digital Finance

Introduction to Part I

The journey through the cryptoverse is an interesting one. On the one hand, it is something new and exciting. The digital world has a frontier like no other, and with that comes a sense of adventure as we discover new things in the system. On the other hand, we see some of the old TradFi systems starting to take shape within the cryptoverse. Elements which we know, which give a familiar feel to the area- even though they are being used in new ways.. In this exploration, we delve into the realms of the cryptoverse, navigate the intricacies of blockchain, and unravel the distinctions between coins and tokens. This adventure takes us through the spiderweb bridge of on-ramps and off-ramps, exploring how Fintech connects traditional finance with the digital frontier.

TradFi- The traditional system of finance.

The cryptoverse, a term encapsulating the expansive realm of cryptocurrencies and blockchain technology, serves as the starting point for our exploration. It is a domain marked by innovation, disruption, and transformative potential. As we traverse the cryptoverse, we encounter the fundamental building blocks – blockchain and cryptocurrency – that lay the groundwork for the broader Fintech landscape.

Our exploration begins with an understanding of blockchain, the distributed ledger technology that underpins cryptocurrencies and facilitates decentralized transactions. The concept of block arrays or block matrixes adds depth to our comprehension, illustrating how data is organized and secured within the blockchain structure. Through this, we gain insight into the foundational technology shaping the future of financial transactions.

Our journey progresses to the heart of the cryptoverse – cryptocurrency. Originating from the marriage of cryptographic principles and decentralized networks, cryptocurrencies offer a paradigm shift from traditional fiat currencies. The evolution of digital assets, their functionalities, and the underlying technologies shaping them become integral aspects of our exploration.

Coins and tokens emerge as distinctive entities within the cryptocurrency realm, each with its unique purpose and utility. This distinction forms a pivotal chapter in our adventure, as we unravel the complexities of native utility tokens and their roles within blockchain ecosystems. The narrative unfolds against the backdrop of the spiderweb bridge, symbolizing the interconnected on-ramps and off-ramps facilitating entry and exit points to and from the cryptoverse.

The exploration extends to permissionless systems, where the traditional constraints of the financial world are transcended. We navigate the uncharted territories of zero-trust systems, witnessing how trust is redefined in the cryptoverse. As we traverse these innovative landscapes, the metaphor of the Wild West comes to life, depicting the untamed nature of Fintech's current state, awaiting the order that regulations may bring.

The journey encapsulates the Fintech ecosystem's multifaceted nature, encompassing digital banking, blockchain, insurtech, and robo-advisors. This landscape is a convergence of traditional and digital finance, paving the way for collaborative partnerships and the concept of open banking.

As we set the stage for the chapters ahead, the Fintech adventure promises a compelling narrative of Web3 mechanics, cross-border payments, and the challenges faced by businesses embracing Fintech. This exploration seeks to demystify the complexities of the financial technology space, offering a panoramic view of the transformative forces shaping the future of finance. The subsequent chapters are poised to deepen our understanding, unveiling new facets of the Fintech ecosystem and steering us towards the frontiers of digital finance.

Chapter 1: Introduction to the Cryptoverse

"If you don't have time to believe it or get it, I don't have time to try to convince you sorry."

Satoshi Nakamoto[1]

"Damnit, I guess that leaves it to the rest of the industry to explain it."

Everyone Else

Introduction:

The cryptoverse, an amalgamation of "crypto" and "universe," represents the vast and interconnected world of cryptocurrencies and blockchain technology. It's an ecosystem that has evolved significantly since its inception. This exploration looks into the definition, historical roots, use cases, and challenges faced by the cryptoverse.

Definition of the Cryptoverse:

The cryptoverse is a virtual realm encompassing all things related to cryptocurrencies and blockchain technology. At its core, it involves a decentralized and secure network of digital assets and transactions.[2] This realm is not confined to a single cryptocurrency but extends to the multitude of blockchain-based innovations that have emerged over time. It also includes the intricacies of Web3 and Fintech as we see the merging of the physical world and the digital world.

Cryptocurrencies, the primary inhabitants of the cryptoverse, are digital or virtual currencies that use cryptography for security and operate on decentralized networks based on

---

[1] Nakamoto, Satoshi (N/A) in "Quotes on Crypto." Trality at https://www.trality.com/blog/the-most-famous-crypto-quotes.
[2] Staff (2023). "The Cryptoverse, more than Just digital Money." Brickken at https://www.brickken.com/post/blog-cryptoverse.

blockchain technology.[3] Blockchain, in essence, is a distributed ledger that records transactions across a network of computers, providing transparency, immutability, and security.

History of the Cryptoverse:

The roots of the cryptoverse can be traced back to the introduction of Bitcoin in 2009 by an individual or group using the pseudonym Satoshi Nakamoto. Bitcoin, the pioneer cryptocurrency, aimed to create a decentralized and peer-to-peer electronic cash system. Its success paved the way for the development of numerous alternative cryptocurrencies (altcoins) and laid the foundation for the broader cryptoverse.[4]

Over the years, the cryptoverse has witnessed significant milestones, including the introduction of smart contracts by Ethereum, enabling programmable and self-executing agreements. The industry has experienced both moments of exuberant growth and periods of regulatory scrutiny. Despite challenges, the cryptoverse has continuously evolved, giving rise to innovative projects and technologies.

Use Cases for the Cryptoverse:

The cryptoverse offers a myriad of use cases that extend beyond traditional financial transactions. Cryptocurrencies serve as a medium of exchange, store of value, and unit of account. Blockchain technology, the backbone of the cryptoverse, finds applications in various industries, such as supply chain, healthcare, and voting systems.

Smart contracts, coded agreements that automatically execute and enforce terms, have revolutionized processes in sectors like

---

[3] *Id.*

[4] Evans, Roy (2023). "The intriguing story of the cryptoverse." Ron Writer at https://www.linkedin.com/pulse/intriguing-story-cryptoverse-ron-evans-wumkc/?trk=article-ssr-frontend-pulse_more-articles_related-content-card.

legal, insurance, and real estate. Decentralized Finance (DeFi) platforms provide financial services without traditional intermediaries, offering opportunities for lending, borrowing, and trading.[5]

The cryptoverse has also become a hub for innovation, with non-fungible tokens (NFTs) gaining prominence. NFTs represent ownership or proof of authenticity for digital or physical assets, opening avenues in art, gaming, and entertainment.

Challenges Faced by the Cryptoverse:

While the cryptoverse holds immense potential, it is not without challenges. Regulatory uncertainties, security concerns, and scalability issues are among the key hurdles. Regulatory bodies globally are grappling with how to classify and regulate cryptocurrencies, impacting markets.[6]

Security breaches and hacking incidents pose threats to the integrity of the cryptoverse. The anonymity associated with some cryptocurrencies raises concerns about their potential use in illicit activities. Scalability challenges, particularly in blockchain networks, affect transaction speed and cost.

Conclusion:

In conclusion, the cryptoverse represents a and expansive realm where cryptocurrencies and blockchain technology converge. Its journey from the inception of Bitcoin to the emergence of various use cases showcases the adaptability and resilience of this evolving ecosystem. While facing challenges, the cryptoverse continues to push boundaries and redefine

---

[5] Brickken, Supra .

[6] Because this book is laid out as a handbook, regulatory compliance will be an issue which crops up in almost every chapter. This is so that those who are reading the chapters out of order have a context for the ways in which the system functions.

possibilities, leaving an indelible mark on the future of finance and technology.

Chapter 2: The Building Blocks- What is Blockchain?

Introduction:

Blockchain, a revolutionary technology underpinning the cryptoverse, stands as a testament to the evolution of decentralized systems. This chapter delves into the core concepts of blockchain, exploring its definition, historical roots, and the fascinating structure of block arrays or block matrixes.

What is a Blockchain:

Blockchain: a peer-to-peer network of computers which record transactions and data transfers.

At its essence, a blockchain is a decentralized and distributed ledger that records transactions across a network of computers in a secure and transparent manner. The term "blockchain" is derived from the way data is structured — transactions are grouped into blocks, and each block is linked to the previous one, forming a continuous chain. This design ensures the immutability and integrity of the recorded information.[7]

The blockchain operates on a consensus mechanism, wherein a network of nodes validates and agrees on the state of the ledger. Transactions, once added to a block, become a permanent part of the blockchain, creating a tamper-resistant and verifiable record. The decentralized nature of blockchain eliminates the need for a central authority, providing trust and transparency in a trustless environment.

---

[7] Hayes, *infra.*

History of Blockchain:

The roots of blockchain trace back to the conceptualization of a cryptographically secure chain of blocks by Stuart Haber and W. Scott Stornetta in 1991.[8] However, the breakthrough came with the introduction of Bitcoin's whitepaper by the mysterious Satoshi Nakamoto in 2008.[9] Nakamoto proposed a peer-to-peer electronic cash system that utilized a blockchain to record and verify transactions without the need for a central authority.

Bitcoin, the first and most well-known application of blockchain, became operational in 2009. Its success sparked interest in the broader potential of blockchain technology beyond digital currencies. Ethereum, introduced in 2015 by Vitalik Buterin, expanded the capabilities of blockchain by introducing smart contracts — self-executing contracts with the terms of the agreement directly written into code.[10]

The years that followed saw the emergence of various blockchain projects exploring applications in supply chain management, healthcare, finance, and more. The technology's decentralized and transparent nature garnered attention from industries seeking to enhance security, efficiency, and trust in their processes.

Block Arrays or Block Matrixes:

One of the intriguing aspects of blockchain technology is its structure of block arrays or block matrixes. In a blockchain, transactions are bundled into blocks, and each block contains a unique identifier called a hash, which is generated using

---

[8] Habar, Stuart & Stornetta, Scott (1991). "How to time stamp a digital document. Journal of Cryptology at https://link.springer.com/article/10.1007/BF00196791.

[9] ICAEW at https://www.icaew.com/technical/technology/blockchain-and-cryptoassets/blockchain-articles/what-is-blockchain/history.

[10] *Id.*

cryptographic algorithms.[11] The hash of each block also includes the hash of the previous block, forming a chain.

The concept of block arrays extends this structure into a matrix-like arrangement, where multiple chains or sequences of blocks run parallelly. Each chain represents a set of transactions or data, and the interconnectedness of these chains adds another layer of complexity to the overall structure. This design enhances scalability and allows for the organization of diverse types of data within the blockchain.

Block matrixes offer flexibility and adaptability, allowing for the categorization of information into different branches or threads. This architecture is particularly beneficial in applications where various types of data need to be managed simultaneously without compromising the integrity of the overall blockchain structure.

Conclusion:

In conclusion, blockchain represents a transformative technology with a rich history and a unique structure of block arrays or block matrixes. From its conceptual origins to the decentralized revolution led by Bitcoin and Ethereum, blockchain has evolved into a foundational element of the cryptoverse. Understanding its structure, especially the concept of block arrays, opens the door to exploring the diverse applications and potential innovations that this revolutionary technology offers.

---

[11] *See Generally* Highman, Nick (2021). "What is a Block Matrix." https://nhigham.com/2020/09/22/what-is-a-block-matrix/. This is a discussion of the function, not a cryptography element, it just gives way to the concept which can be forthcoming.

Chapter 3: Currency of the Digital Realm- What is Cryptocurrency?

Introduction:

The inception and evolution of cryptocurrency have reshaped the landscape of finance, introducing a decentralized and secure means of transacting value. This chapter explores the origin and functionality of cryptocurrency, delving into the transformative impact it has had on traditional financial systems.

Cryptocurrency Origins:

Cryptocurrency: digital currency on the blockchain where each transaction is recorded by a decentralized system of cryptography via a ledger.

The roots of cryptocurrency trace back to the global financial crisis of 2008. An individual or group using the pseudonym Satoshi Nakamoto presented the concept in a whitepaper titled "Bitcoin: A Peer-to-Peer Electronic Cash System." Released in 2008, the whitepaper outlined the design of a decentralized digital currency that would operate without the need for a central authority.[12]

In 2009, Nakamoto implemented the first cryptocurrency, Bitcoin, as an open-source software. Bitcoin's underlying technology, blockchain, served as a distributed ledger that recorded and verified transactions through a network of nodes. This innovative approach aimed to address the inefficiencies and vulnerabilities inherent in traditional banking systems.[13]

---

[12] Frankenfield, Jake (2023). "Cryptocurrency Explained with Pros and Cons for Investment." Investopedia at https://www.investopedia.com/terms/c/cryptocurrency.asp.
[13] *Id.*

Functionality of Cryptocurrency:

Cryptocurrencies operate on a set of core principles that distinguish them from traditional forms of money:

1. **Decentralization:** Cryptocurrencies eliminate the need for central authorities, such as banks or governments, by employing a decentralized network of nodes to validate and record transactions. This decentralized structure enhances security and prevents a single point of failure.
2. **Blockchain Technology:** The backbone of most cryptocurrencies is blockchain, a distributed ledger that maintains a secure and transparent record of all transactions. Each block in the chain contains a hash of the previous block, ensuring the integrity and immutability of the entire transaction history.
3. **Cryptography:** Cryptocurrencies rely on cryptographic techniques to secure transactions and control the creation of new units. Public and private keys, along with cryptographic hashing, play a crucial role in ensuring the authenticity and privacy of users' transactions.
4. **Limited Supply:** Many cryptocurrencies, including Bitcoin, have a capped supply, creating scarcity similar to precious metals like gold. This limited supply is programmed into the underlying code, preventing inflationary pressures often associated with traditional fiat currencies.
5. **Anonymity and Pseudonymity:** While transactions are recorded on the blockchain, users' identities are often protected through pseudonymous addresses. This provides a level of privacy, although the degree varies among different cryptocurrencies.
6. **Global Accessibility:** Cryptocurrencies transcend geographical boundaries, allowing anyone with an

internet connection to participate in the network. This inclusivity is particularly impactful in regions with limited access to traditional banking services.

7. **Smart Contracts:** Some cryptocurrencies, like Ethereum, introduced the concept of smart contracts. These self-executing contracts with coded terms automate and enforce contractual agreements, eliminating the need for intermediaries in various applications.

Conclusion:

In conclusion, cryptocurrency emerged as a disruptive force challenging conventional notions of money and finance. Born out of a desire for financial autonomy and transparency, cryptocurrencies like Bitcoin introduced a novel approach to value exchange. Their functionality, rooted in decentralization, blockchain technology, cryptography, and global accessibility, has paved the way for a new era in finance. As we explore the world of cryptocurrencies, the next chapter will delve into the distinctions between coins and tokens, providing a nuanced understanding of these digital assets' diverse functionalities.

Chapter 4: Navigating the Basics- Coins v. Tokens

Coin: The native utility token on a blockchain.

Token: Any cryptocurrency operating on a blockchain, either natively or through a smart contract.

Introduction:

Within the expansive realm of cryptocurrencies, the terms "coin" and "token" are frequently used, often interchangeably, leading to confusion among enthusiasts and newcomers alike. This chapter seeks to elucidate the distinctions between coins and tokens in the cryptocurrency landscape, shedding light on their unique characteristics, use cases, and underlying technologies.

Coins vs. Tokens: Unveiling the Differences

- Definition:
    - Coins: Coins are native digital currencies that operate on their blockchain. Bitcoin, the pioneer cryptocurrency, is a prime example of a coin. These coins serve as a medium of exchange, store of value, and unit of account within their respective blockchain networks.[14]
    - Tokens: Tokens, on the other hand, represent a broader category of digital assets built on existing blockchain platforms. They are not stand-alone currencies but rather derive their utility and value from the functionalities of the

[14] Smithmyer *et al* (2022), *infra*.

platforms they inhabit, such as Ethereum or Binance Smart Chain.[15]

- Blockchain Origin:
    - Coins: Coins are typically the original native assets of a blockchain. For instance, Ether (ETH) is the native coin of the Ethereum blockchain, facilitating transactions and fueling smart contracts.
    - Tokens: Tokens are created through a process known as Initial Coin Offerings (ICOs) or Token Generation Events (TGEs) on existing blockchain platforms. They are, in essence, a representation of assets or utility within a specific project or decentralized application (DApp).[16]
- Purpose and Functionality:
    - Coins: The primary purpose of coins is to function as a medium of exchange. Users can send and receive coins, and their values are influenced by factors such as scarcity, demand, and overall network adoption.
    - Tokens: Tokens, while often used as a medium of exchange, have a broader range of functionalities. They can represent ownership in a decentralized application, grant access to specific features, or even symbolize real-world assets like real estate or commodities.
- Use Cases:
    - Coins: Coins are generally used for transactions and as an investment or speculative asset. Bitcoin, for example, is often held as a digital store of value similar to precious metals.

---

[15] *Id.*

[16] *Id.*

    - Tokens: Tokens serve diverse use cases. They can represent ownership in decentralized finance (DeFi) protocols, provide governance rights, enable access to a particular platform, or even embody the value of physical assets through tokenization.
- Platform Dependency:
    - Coins: Coins are not dependent on any other platform. They exist on their native blockchain and operate independently.
    - Tokens: Tokens are intrinsically tied to the blockchain platform on which they are created. For example, ERC-20 tokens exist on the Ethereum blockchain, while BEP-20 tokens reside on the Binance Smart Chain and SPL/Token2022 Tokens reside on the Solana Blockchain.

Conclusion:

In conclusion, understanding the distinction between coins and tokens is crucial for navigating the landscape of cryptocurrencies. Coins operate as native currencies on their respective blockchains, serving fundamental roles in transactions and value storage. In contrast, tokens leverage existing blockchain infrastructure to represent a spectrum of assets, utilities, and functionalities within specific projects or ecosystems.

This nuanced understanding empowers investors, developers, and enthusiasts to make informed decisions based on the unique attributes and applications of coins and tokens. As the cryptocurrency ecosystem continues to evolve, these distinctions become increasingly relevant, shaping the diverse and multifaceted nature of digital assets in the broader financial landscape.

Chapter 5: On-Ramps and Off-Ramps- Traveling the Spiderweb Bridge in and out of the Cryptoverse

Introduction:

Spiderweb Bridge: The system of interconnectedness within the cryptoverse.

The world of cryptocurrency operates as a vast network interconnected by various entry and exit points, often referred to as on-ramps and off-ramps. Understanding these access points is akin to navigating a spiderweb bridge, connecting traditional financial systems with the decentralized landscape of the cryptoverse. In this chapter, we delve into the significance of on-ramps and off-ramps, exploring their functionalities, impact on user experiences, and their crucial role in bridging the gap between fiat currencies and the digital realm of cryptocurrencies.

On-Ramps: Gateways to the Cryptoverse

On-ramps represent the entry points for individuals seeking to convert fiat currency into cryptocurrencies. These gateways facilitate the transition from traditional financial systems to the decentralized realm of blockchain and digital assets. On-ramps play a pivotal role in introducing newcomers to the cryptoverse, acting as bridges between conventional banking and the emerging landscape of cryptocurrency.[17]

- Types of On-Ramps:
    - *Credit Card Purchases:* Many cryptocurrency exchanges and platforms allow users to purchase digital assets using credit or debit

---

[17] Alchemy (2024) at https://www.alchemy.com/best/fiat-onramps.

cards, providing a straightforward and familiar on-ramp.
  - *Bank Transfers:* On-ramps also include platforms that enable users to transfer funds directly from their bank accounts to purchase cryptocurrencies, promoting a seamless integration with traditional banking systems.
  - *Peer-to-Peer Platforms:* Certain decentralized platforms facilitate peer-to-peer transactions, acting as on-ramps where users can buy and sell cryptocurrencies directly with one another.
- Key Considerations for On-Ramps:
  - *Fees:* Different on-ramps may impose varying fees for transactions. It is essential for users to consider these costs when selecting an on-ramp based on their preferences and financial considerations.
  - *Convenience:* The ease and speed of on-ramp transactions are critical factors influencing user experiences. Platforms that offer a user-friendly interface and swift transaction processing enhance accessibility for a broader audience.
  - *Transaction Limits:* On-ramps may have specific limits on transaction amounts. Understanding these limits is crucial for users planning to invest or transact with larger sums.
  - *Supported Currencies:* Users should consider on-ramps that support their desired fiat currencies, ensuring compatibility and accessibility based on their geographic location.

Off-Ramps: Cashing Out from the Cryptoverse

Off-ramps, conversely, facilitate the conversion of cryptocurrencies back into fiat currency, enabling users to withdraw and utilize their digital assets in traditional financial systems. These exit points play a crucial role in providing liquidity and ensuring that users can seamlessly transition between the cryptoverse and traditional financial structures.[18]

- **Types of Off-Ramps:**
    - *Exchange Withdrawals:* Users can convert their cryptocurrencies into fiat currency directly on cryptocurrency exchanges, facilitating withdrawals to bank accounts linked to the exchange.
    - *P2P Platforms:* Similar to on-ramps, certain peer-to-peer platforms enable users to sell their cryptocurrencies directly to other individuals, receiving fiat currency in return.
    - *Crypto ATMs:* Off-ramps can also take the form of cryptocurrency ATMs, allowing users to withdraw fiat currency by selling their digital assets through these automated machines.
- **Key Considerations for Off-Ramps:**
    - *Fees:* Off-ramps may impose transaction fees and withdrawal charges. Users should be mindful of these costs to optimize the conversion of cryptocurrencies into fiat currency.
    - *Speed of Transaction:* The speed at which off-ramps process transactions is a crucial consideration, especially for users looking to quickly access funds in traditional financial systems.
    - *Conversion Rates:* Understanding the conversion rates offered by different off-ramps is essential,

---

[18] Staff (2024). "Crypto Ramps and our Web3 Future." Kriptomat at https://kriptomat.io/best-fiat-to-crypto-on-ramp/what-are-cryptocurrency-ramps-and-web3-future/.

as it directly influences the amount of fiat currency users receive when cashing out their cryptocurrencies.

- *Supported Fiat Currencies:* Off-ramps may have limitations on the fiat currencies they support. Users should choose off-ramps that align with their preferred currency options.

Conclusion:

In conclusion, on-ramps and off-ramps serve as the vital connectors between the cryptoverse and traditional financial systems, forming the spiderweb bridge that enables seamless transitions for cryptocurrency users. These access points not only facilitate the entry and exit of individuals into and from the world of digital assets but also contribute to the overall liquidity and accessibility of the cryptoverse. As the ecosystem continues to evolve, the efficiency and reliability of on-ramps and off-ramps will play a pivotal role in shaping the future landscape of cryptocurrency adoption and integration into mainstream financial practices.

Chapter 6: Permissionless Systems- Breaking Free From Traditional Constraints

Introduction:

In the relentless evolution of financial systems, permissionless systems stand as a revolutionary force, challenging the traditional constraints imposed by the Traditional Financial (TradFi) world.[19] At the core of the cryptoverse, these permissionless systems redefine access, empowerment, and autonomy for users. In this exploration, we delve into the profound impact of permissionless systems, their principles, and how they liberate financial interactions from the conventions of the traditional banking sphere.

Understanding Permissionless Systems:

1. **Decentralization and Autonomy:** Permissionless systems operate on the principles of decentralization, removing the need for a central authority to control or validate transactions.[20] Users gain autonomy over their financial activities, bypassing intermediaries and enjoying direct peer-to-peer interactions.
2. **Inclusive Access:** Unlike traditional financial systems that often impose barriers to entry, permissionless systems are inclusive by design. Anyone with an internet connection can participate, transcending geographical

---

[19] Staff (2022). "What does permissionless mean." Cryptopedia at https://www.gemini.com/cryptopedia/what-is-permissionless-crypto-permissionlessness-blockchain.

[20] Groopman, Jessica (2023). "Permissioned v. Permissionless Blockchains: Key Differences." Tech Target at https://www.techtarget.com/searchcio/tip/Permissioned-vs-permissionless-blockchains-Key-differences.

boundaries and offering financial services to the unbanked and underserved populations globally.[21]

3. **Blockchain Technology as the Enabler:** Permissionless systems leverage blockchain technology as the underlying infrastructure. The blockchain serves as an immutable and transparent ledger, ensuring the integrity of transactions without the need for centralized oversight.[22] Smart contracts further automate and enforce agreements without relying on traditional legal frameworks.

Breaking Free from TradFi Constraints:

1. **Financial Inclusion:** Permissionless systems champion financial inclusion by providing access to banking services for individuals who have been historically excluded.[23] The decentralized nature of these systems ensures that anyone with a smartphone and internet connectivity can participate in the financial ecosystem.
2. **Eliminating Intermediaries:** Traditional financial systems often involve multiple intermediaries, each adding a layer of complexity and cost to transactions. Permissionless systems streamline processes by eliminating unnecessary middlemen, reducing fees, and enhancing the speed of financial interactions.[24]

---

[21] Staff (2024). "Permissioned and Permissionless Blockchains: Differences and Benefits." Ediie at https://www.ediiie.com/blog/permissioned-and-permissionless-blockchains/.

[22] Groopman, *Supra*.

[23] Staff (2022). "Blockchain: Financial and Non-Fiancial Uses and Challenges." GAO at https://www.gao.gov/blog/blockchain-financial-and-non-financial-uses-and-challenges.

[24] Reiff, Nathan (2021). "Blockchain won't cut out intermediaries after all." Investopedia at https://www.investopedia.com/tech/blockchain-wont-cut-out-intermediaries-after-all/.

3. **Global Accessibility:** Permissionless systems operate on a global scale, offering financial services to individuals regardless of their location.[25] This global accessibility is a stark departure from the limitations imposed by traditional banking, especially in regions with underdeveloped financial infrastructure.

Challenges and Considerations:

1. **Regulatory Uncertainty:** The permissionless nature of these systems poses challenges in the realm of regulatory compliance. Traditional financial institutions, bound by stringent regulations, grapple with the decentralized and pseudonymous aspects of permissionless systems.
2. **Volatility and Risk:** The inherent volatility of cryptocurrencies, often associated with permissionless systems, can be a double-edged sword. While it presents opportunities for rapid gains, it also introduces higher levels of risk, contrasting with the relative stability found in traditional financial assets.

Conclusion:

In conclusion, permissionless systems epitomize the spirit of decentralization and empowerment within the cryptoverse. They serve as a disruptive force, challenging the status quo of traditional financial systems and redefining the landscape of global finance. The inclusive, borderless, and efficient nature of permissionless systems not only addresses the shortcomings of traditional banking but also opens new frontiers for financial innovation and accessibility. As the cryptoverse continues to mature, the integration of permissionless systems signifies a

---

[25] Staff (2023). "Understanding Permissioned and Permissionless Blockchains." Super EX Nigeria at https://medium.com/@superexnigeria/understanding-permissioned-and-permissionless-blockchains-8b8b76860542.

paradigm shift, offering users a decentralized future where financial control is placed firmly in their hands.

Chapter 7: Zero-Trust Systems- Rethinking Trust in Transactions

Introduction:

Zero Trust Systems: a system which requires identity verification from any users within the system.

In the realm of the cryptoverse, where security and trust are paramount concerns, the concept of zero trust systems emerges as a robust safeguard against potential threats. Unlike traditional security models that operate on a perimeter-based trust approach, zero trust systems assume that threats can emerge from within and outside the network.[26] This exploration delves into the principles, mechanisms, and significance of zero trust systems in the cryptoverse, elucidating their pivotal role in fortifying the security architecture.[27]

Understanding Zero Trust Systems:

1. **Fundamental Principles:** Zero trust systems operate on the fundamental principle of "never trust, always verify." In essence, no entity, whether inside or outside the network, is granted implicit trust. Instead, continuous verification and authentication are required for every user, device, or application attempting to access resources.[28]
2. **Micro-Segmentation:** A core tenet of zero trust is micro-segmentation, which involves dividing the network into

[26] Raina, Kapil (2023). "Zero Trust Security Explained: Principles of the Zero Trust Model." Crowdstrike at https://www.crowdstrike.com/cybersecurity-101/zero-trust-security/.
[27] *Id.*
[28] *Id.*

smaller, isolated segments.[29] Each segment has its own set of access controls, and communication between segments is strictly monitored. This granular approach limits the lateral movement of potential threats within the network.

3. **Authentication:** Traditional models rely on static authentication measures, often requiring users to authenticate only at the initial login. In contrast, zero trust systems enforce authentication, necessitating continuous verification throughout a user's session.[30] This adaptability enhances security by responding to real-time changes in user behavior.

Operational Mechanisms of Zero Trust in the Cryptoverse:

1. **Blockchain as a Trust Anchor:** In the cryptoverse, where decentralization is a foundational principle, blockchain technology serves as a trust anchor within zero trust systems.[31] The immutable and transparent nature of the blockchain ensures the integrity of transactions and validates the authenticity of participants.
2. **Smart Contracts and Access Control:** Smart contracts, programmable self-executing agreements, play a pivotal

---

[29] Staff (2023). "Micro-Segmentation: The First Step to Zero-Trust Security." Color Tokens at https://colortokens.com/blog/micro-segmentation-first-step-zero-trust-security/.

[30] Yao, Qigui; Wang, Qi; Zhang, Xiaojian; & Fei, Jiaxuan (2020). " Access Control and Authorization Systems Based on Zero Trust Architecture." ACM at https://dl.acm.org/doi/abs/10.1145/3437802.3437824.

[31] King, Alan; Williams, Chris; Herching, Aliza; Moshkovich, Dany; Tock, Yoav (2022). "Trust Anchor: Zero Trust Architecture for Digital Payments." SSRN at https://papers.ssrn.com/sol3/papers.cfm?abstract_id=4059570.

role in zero trust systems.[32] They enable automated and decentralized access control, ensuring that only authorized entities can execute predefined actions within the blockchain network.

3. **Decentralized Identity Management:** Traditional identity management systems are often centralized, making them susceptible to single points of failure.[33] Zero trust systems in the cryptoverse leverage decentralized identity solutions, such as decentralized identifiers (DIDs) and verifiable credentials, to empower users with greater control over their digital identities.

Significance in the Cryptoverse:

1. **Enhanced Security Posture:** Zero trust systems significantly enhance the security posture of blockchain networks and cryptocurrency platforms. By eliminating the reliance on assumed trust within the network, these systems reduce the attack surface and mitigate the risk of unauthorized access.[34]
2. **Protection Against Insider Threats:** Cryptocurrency ecosystems are not immune to insider threats.[35] Zero trust systems, with their stringent verification mechanisms, offer robust protection against malicious activities originating from within the network, ensuring

---

[32] Gai, Keke; She, Yufeng; Zhu, Liehuang; Choo, Raymond & Wan, Zhiguo (2023). "A Blockchain Based Access Control Scheme for Zero Trust Cross Organizational Data Sharing." ACM at https://dl.acm.org/doi/abs/10.1145/3511899.

[33] Staff (2023). "Zero Trust Security- What is a Zero Trust Network." Cloude Flare at https://www.cloudflare.com/learning/security/glossary/what-is-zero-trust/.

[34] Raina, *supra*.

[35] Dunn, Peter (2022). "Insiders may be the biggest zero-trust threat." FedTech at https://fedtechmagazine.com/article/2022/10/insiders-may-be-biggest-zero-trust-threat.

that even authorized users undergo continuous scrutiny.

Challenges and Considerations:

1. **Scalability and Performance:** Implementing zero trust systems in the cryptoverse requires careful consideration of scalability and performance. The cryptographic processes and continuous authentication mechanisms may introduce latency, impacting the overall efficiency of the network.
2. **User Experience:** Striking a balance between heightened security measures and a seamless user experience poses a challenge. Ensuring that the stringent verification processes do not impede the usability of cryptocurrency platforms is a crucial consideration.

Conclusion:

In conclusion, zero trust systems represent a paradigm shift in cybersecurity within the cryptoverse, aligning with the principles of decentralization and continuous verification. By adopting a and granular approach to access controls, these systems fortify the security architecture of blockchain networks and cryptocurrency platforms. The incorporation of blockchain technology, smart contracts, and decentralized identity solutions underscores the unique adaptability of zero trust systems in the context of the cryptoverse. As the cryptoverse continues to evolve, the role of zero trust systems becomes increasingly pivotal in ensuring the integrity, confidentiality, and availability of digital assets and transactions.

Chapter 8: Exploring Decentralization- Redefining the Nation State

Introduction:

Decentralization stands as a foundational principle within the cryptoverse, embodying a paradigm shift away from traditional centralized systems. This exploration delves into the concept of decentralization, unraveling its essence, its role in democratizing systems, and the implications for the traditional concept of the nation state.

Understanding Decentralization:

> *Decentralized Finance: a system based on secure distributed ledgers rather than individual centralized authorities.*

1. **Democratizing Systems:** At its core, decentralization is a mechanism that disperses control and authority across a network, eschewing a single central authority. In doing so, it democratizes systems, fostering a more inclusive and participatory environment. In the cryptoverse, decentralized networks are often governed by consensus mechanisms, where decisions are made collectively by participants rather than a centralized entity.
2. **Distributed Ledger Technology:** The advent of blockchain, a decentralized and distributed ledger technology, exemplifies the application of decentralization in the cryptoverse. Transactions are recorded across a network of nodes, eliminating the need for a central authority, enhancing transparency, and mitigating the risk of a single point of failure.

Decentralization and Democracy:

1. **Empowering Individuals:** Decentralization empowers individuals within the cryptoverse by providing them with direct control over their digital assets. Users, rather than relying on intermediaries, become custodians of their private keys and, consequently, their financial sovereignty. This empowerment aligns with democratic principles, as individuals have a greater say and ownership in the financial systems they participate in.
2. **Inclusive Decision-Making:** Consensus mechanisms, such as Proof of Stake (PoS) or Proof of Work (PoW), ensure that decisions regarding the protocol or network upgrades are made collectively.[36] This inclusive decision-making process, open to all participants, mirrors democratic ideals where stakeholders have a voice in the governance of the system.

Decentralization and the Nation State:

1. **Challenges to Traditional Governance:** The concept of decentralization poses challenges to the traditional notion of the nation state as the primary authority. In a decentralized landscape, governance is distributed, making it challenging for a centralized entity to exert control. Cryptocurrencies, operating on decentralized networks, are not bound by national borders, challenging the traditional regulatory authority of nation states.
2. **Global Nature of Cryptocurrencies:** Cryptocurrencies, being decentralized and borderless, transcend the jurisdictional boundaries traditionally defined by nation states. This global nature introduces complexities for regulatory frameworks, taxation, and enforcement. The decentralized nature of blockchain networks challenges

---

[36] Visa, *infra*.

the conventional role of the nation state as the sole arbiter of financial governance.

Challenges and Considerations:

1. **Regulatory Uncertainty:** The decentralization of cryptocurrencies has prompted regulatory uncertainty, with nation states grappling to formulate cohesive policies. Striking a balance between innovation and regulatory control remains a challenge for governments seeking to adapt to the changing landscape.
2. **Identity and Accountability:** The decentralized nature of blockchain networks often challenges traditional notions of identity and accountability. As transactions occur pseudonymously, establishing a framework for legal recourse and responsibility becomes a nuanced consideration.

Conclusion:

In conclusion, decentralization emerges as a transformative force within the cryptoverse, redefining power s and challenging traditional notions of governance. Its role in democratizing systems aligns with principles of inclusivity and participatory decision-making. However, the implications for the nation state are profound, as decentralized networks introduce complexities that demand a reevaluation of regulatory frameworks. Striking a balance between the advantages of decentralization and the need for regulatory oversight remains an ongoing challenge. As the cryptoverse continues to evolve, the interplay between decentralization and the nation state will undoubtedly shape the future trajectory of both financial systems and governance structures.

Chapter 9: Fintech Wild West- Regulatory Challenges and Opportunities

Introduction:

In the vast and landscape of financial technology, commonly known as Fintech, one can't help but draw parallels to the Wild West—a realm of untamed innovation and uncharted territories. As businesses and entrepreneurs explore the potential of Fintech, the metaphor of the Wild West captures the spirit of adventure, risk-taking, and the absence of a fully established regulatory framework. However, just like the Wild West saw the arrival of law and order to establish stability, the Fintech industry is witnessing the inevitable evolution toward regulatory oversight.

The Fintech Frontier:

> *Fintech: a system of banking and finance programs designed to enable banking and financial services.*

Fintech, characterized by the convergence of finance and technology, has become a breeding ground for groundbreaking ideas, disruptive technologies, and innovative financial services. Similar to the Wild West, where pioneers sought fortune and opportunity in unexplored territories, Fintech enthusiasts are venturing into new realms—cryptocurrencies, blockchain, decentralized finance (DeFi), and more.[37] This unbridled exploration has led to a vibrant ecosystem, fostering creativity and pushing the boundaries of traditional finance.

---

[37] Staff (2023). "Unveiling the Fintech Frontier: Your Guide to the Future of Finance." Fintech IO at https://www.linkedin.com/pulse/unveiling-fintech-frontier-your-guide-future-finance-fintechio/.

As the Fintech landscape matures, regulators are beginning to stake their claim on this digital frontier. The absence of a comprehensive regulatory framework has allowed Fintech to flourish, but it has also given rise to concerns such as fraud, security risks, and potential market abuses. Governments and regulatory bodies worldwide are recognizing the need to establish rules and oversight to ensure the stability and security of the financial system.[38]

The Inevitable Convergence:

The arrival of regulations in the Fintech space is akin to the establishment of law and order in the Wild West. It signifies a shift from the lawless, frontier-like environment to a more structured and secure landscape.[39] Regulatory measures are being introduced to address challenges such as customer protection, financial stability, and the prevention of illicit activities. While some view these regulations as constraints, they are essential for the sustainable growth and mainstream acceptance of Fintech.

For Fintech businesses, adapting to regulatory changes is both a challenge and an opportunity. Navigating compliance requirements demonstrates a commitment to ethical practices and consumer protection. It allows businesses to build trust with users, investors, and regulatory authorities, paving the way for long-term success. The shift toward regulatory compliance also opens doors to broader market acceptance and collaboration with traditional financial institutions. Regulations serve as a guidepost for ethical conduct, defining the

---

[38] *Id.*

[39] Mercer, David (2023). "TradFi and DeFi: Inevitable Convergence for the Future." Traders Magazine at https://www.tradersmagazine.com/am/tradfi-and-defi-inevitable-convergence-for-the-future/.

boundaries within which Fintech businesses can operate.[40] They provide clarity on permissible activities, ensuring a level playing field for all participants. By establishing regulatory frameworks, governments aim to protect consumers, prevent financial crimes, and maintain the stability of the financial system.

Striking a Balance:

The journey toward a regulated Fintech landscape is about striking a delicate balance between fostering innovation and safeguarding the interests of all stakeholders. While some early adopters of Fintech appreciated the lack of regulatory constraints, the industry's continued growth necessitates a more mature and responsible approach.[41] Regulatory compliance should not be viewed as a hindrance but as a necessary evolution that ensures the industry's credibility and longevity.

Different regions are approaching Fintech regulation with varied perspectives. Some countries are embracing innovation-friendly regulations, fostering the growth of Fintech hubs. Others are adopting a cautious approach, prioritizing consumer protection and risk mitigation. This global diversity in regulatory approaches reflects the complex nature of Fintech and the need for nuanced solutions tailored to each region's unique challenges and opportunities.

Conclusion:

The Fintech frontier may have started as a wild, lawless space, but as with any evolving ecosystem, regulation becomes a necessary pillar for its sustainable development. The regulatory journey is not about stifling innovation but rather about

---

[40] *Id.*

[41] Spinaci, Laura (2022). "TradFi & DeFi Convergence." Medium at https://lauraspinaci.medium.com/the-convergence-between-tradfi-defi-1-2-251268f0bac5.

fostering responsible growth, protecting stakeholders, and establishing a framework for the coexistence of traditional finance and emerging technologies. In this ever-changing landscape, businesses that embrace and adapt to regulatory measures will not only survive but thrive in the new era of Fintech. The frontier may be tamed, but the spirit of innovation and exploration continues to drive the evolution of financial technology.

Chapter 10: The Fintech Ecosystem- Holistic View and Future Horizons

Introduction:

The Fintech ecosystem, a fusion of finance and technology, is undergoing a transformative evolution, reshaping the landscape of traditional financial services. As digital innovation accelerates, Fintech is expanding its reach into various sectors, redefining how individuals and businesses manage, invest, and interact with financial resources. This exploration delves into the current state of the Fintech ecosystem, its key components, and the trajectory it is likely to follow.

Components of the Fintech Ecosystem:

- **Digital Banking:** Fintech's influence on banking is profound, with digital banks emerging as key players. These banks leverage technology to offer seamless, user-friendly experiences, challenging traditional banking models. From mobile banking to virtual wallets, the digital banking segment is thriving.
- **Blockchain and Cryptocurrencies:** The adoption of blockchain technology and cryptocurrencies has disrupted traditional finance, offering decentralized and secure alternatives. Cryptocurrencies like Bitcoin and Ethereum, powered by blockchain, present new paradigms for transactions, investments, and even the concept of money itself.
- **Insurtech:** Insurtech is revolutionizing the insurance sector by harnessing technology for more efficient underwriting, personalized policies, and streamlined claims processes. From AI-driven risk assessments to blockchain-based smart contracts, the marriage of tech and insurance is reshaping the industry.

- **Robo-Advisors:** Fintech-driven robo-advisors are democratizing investment opportunities. These automated platforms use algorithms and AI to provide cost-effective, data-driven investment advice, making wealth management accessible to a broader audience.
- **Peer-to-Peer Lending:** Fintech has disrupted traditional lending models through peer-to-peer lending platforms. These platforms connect borrowers directly with lenders, reducing the need for traditional intermediaries and offering more flexible lending options.
- **Regtech:** Regulatory technology, or Regtech, addresses the increasing complexity of compliance and regulatory requirements. Fintech solutions in this space streamline regulatory processes, enhance transparency, and mitigate risks associated with non-compliance.
- **Embedded Finance:** The concept of embedded finance involves integrating financial services seamlessly into non-financial platforms and applications. From e-commerce platforms offering payment solutions to ride-sharing apps providing insurance, embedded finance enhances user experiences.

Evolution of the Fintech Ecosystem:

The Fintech ecosystem's evolution is marked by a shift from disruption to collaboration. Initially viewed as disruptors challenging traditional financial institutions, Fintech firms are now entering collaborative partnerships with banks and established players. This shift reflects a recognition of the complementary strengths each brings to the table.

1. **Collaborative Partnerships:** Fintech companies and traditional financial institutions are increasingly forming partnerships to create synergies. Banks leverage Fintech's agility and innovation, while Fintech benefits

from the scale and customer base of established financial institutions.

2. **Open Banking:** The concept of open banking, facilitated by Fintech, emphasizes data-sharing between banks and third-party providers. This approach fosters competition, innovation, and the development of new financial products and services.
3. **Global Expansion:** The Fintech revolution is not confined to specific regions. Fintech companies are expanding globally, adapting their solutions to diverse regulatory environments. Cross-border collaborations and expansions are becoming key strategies for Fintech firms seeking to tap into new markets.
4. **Focus on Financial Inclusion:** Fintech is playing a pivotal role in addressing financial inclusion challenges. By leveraging mobile technology and innovative solutions, Fintech aims to provide financial services to underserved populations, bridging gaps in traditional banking infrastructure.
5. **Rise of Decentralized Finance (DeFi):** Decentralized Finance, powered by blockchain, is a disruptive force within the Fintech ecosystem. DeFi projects are reimagining traditional financial services like lending, borrowing, and trading in a decentralized, permissionless environment.

The future trajectory of the Fintech ecosystem is poised for continued innovation and expansion. Several key trends will shape its evolution:

1. **Integration of AI and Machine Learning:** The integration of AI and machine learning will enhance the predictive capabilities of Fintech solutions, offering more personalized and data-driven financial services.

2. **Enhanced Cybersecurity Measures:** As Fintech deals with sensitive financial data, cybersecurity will be paramount. Innovations in cybersecurity technologies will be crucial to building trust and ensuring the secure functioning of the Fintech ecosystem.
3. **Continued Blockchain Adoption:** Blockchain's impact on Fintech will persist, with continued adoption in areas like digital identities, supply chain finance, and the tokenization of assets. Smart contracts and decentralized applications (DApps) will become more prevalent.
4. **Regulatory Evolution:** Regulatory frameworks will continue to evolve to accommodate the complexities of Fintech. Striking a balance between innovation and regulatory oversight will be crucial for the sustained growth of the ecosystem.

Conclusion:

The Fintech ecosystem stands at the forefront of a financial revolution, reshaping the way individuals and businesses engage with financial services. From digital banking to blockchain-powered innovations, the ecosystem reflects a convergence of technology and finance that transcends traditional boundaries. The collaborative spirit between Fintech firms and traditional institutions, coupled with a global expansion mindset, positions the ecosystem for a future defined by innovation, inclusion, and transformative possibilities. As Fintech navigates this ever-changing landscape, its ability to adapt, collaborate, and address emerging challenges will determine its enduring impact on the financial industry.

Conclusion to Part I

In conclusion, the exploration of the Fintech ecosystem in Part I unveils a landscape defined by innovation, collaboration, and transformative potential. As we journeyed through the intricacies of the cryptoverse, looks into blockchain fundamentals, and dissected the distinctions between coins and tokens, a comprehensive understanding of the evolving financial technology space emerged.

The Fintech ecosystem is akin to a adventure, where traditional financial norms meet the uncharted territories of digital finance. This journey introduced us to the spiderweb bridge of on-ramps and off-ramps, facilitating seamless connectivity within the cryptoverse. We navigated the realms of permissionless systems and zero-trust architectures, witnessing how Fintech breaks free from conventional constraints, offering users unprecedented control and security.

The exploration into decentralization revealed a democratization of financial systems, challenging the dominance of traditional nation-states. As we continued our adventure, the Wild West analogy highlighted the current state of Fintech, marked by untamed innovation, disruption, and an impending era of regulations aiming to bring order to this frontier.

The Fintech ecosystem, characterized by its multifaceted components, including digital banking, blockchain, insurtech, and robo-advisors, is shaping a new era of financial services. Collaborative partnerships and the concept of open banking showcase a shift from disruption to synergy, as Fintech firms and traditional institutions join forces.

Looking ahead, the global expansion of Fintech, coupled with a focus on financial inclusion and the rise of decentralized finance (DeFi), points towards an exciting trajectory. The integration of AI and machine learning, continued blockchain adoption, and regulatory evolution are key trends that will define the future of the Fintech ecosystem.

As the Fintech adventure unfolds, the emphasis on enhanced cybersecurity measures becomes paramount, safeguarding sensitive financial data. The chapters underscore the pivotal role played by Fintech in addressing financial inclusion challenges, with innovative solutions designed to reach underserved populations.

In essence, the Fintech ecosystem is not merely a technological evolution; it is a revolution reshaping the global financial landscape. The collaborative spirit, innovative solutions, and a commitment to financial inclusion position Fintech as a transformative force with the potential to redefine how individuals and businesses engage with financial services.

This exploration lays the foundation for the subsequent chapters, which will delve deeper into the intricacies of Web3, cross-border payments, and the challenges and opportunities that businesses face in embracing Fintech. As the journey continues, the Fintech adventure promises to unlock new frontiers, bringing forth a future where digital finance seamlessly integrates with the traditional, creating a financial landscape that is both and inclusive.

Part II: The Blockchain Archipelago- Discovering the Islands of Interoperability

Introduction to Part II

Welcome to the next step of our journey — Part II: Navigating the Cryptoverse. In the previous chapters, we laid the groundwork, introducing you to the basics of blockchain, the essence of decentralization, and the transformative power of cryptocurrency. Now, as we voyage deeper into the cryptoverse, our focus shifts to the web of innovations and applications that define this ecosystem.

Our odyssey commences with an exploration of smart contracts, the digital architects of decentralized agreements. These self-executing contracts, residing on the blockchain, revolutionize the way we conceptualize and enforce contractual obligations. As we delve into their functionality, querying mechanisms, and limitless potential, a new realm of possibilities awaits. We look at the bustling realm of decentralized finance (DeFi), a disruptive force reshaping traditional financial paradigms. Peer into the workings of this financial frontier, where lending, borrowing, and trading occur without intermediaries. The decentralized nature of DeFi opens new vistas for financial inclusion and innovation.

In our pursuit of understanding the nuances within the cryptoverse, we confront the fundamental distinction between coins and token. Then, as we traverse this terrain, it becomes evident that these cryptographic entities play unique roles, each contributing to the broader ecosystem of the digital economy. The exploration then deepens, where we unravel the concept and methodology of tokenization. From transforming supply chain s to revolutionizing identity management, tokens emerge as powerful instruments reshaping industries across the spectrum.

Our journey takes an international turn, where we grapple with the intricacies of cross-border payments. As we embark on this exploration, we anticipate the challenges and opportunities that arise when financial transactions traverse geographical boundaries. We next confront the delicate dance between data security and privacy within the cryptoverse. As technology advances, so does the need for a nuanced understanding of how these elements intersect, ensuring a delicate balance that preserves individual privacy while upholding the security of valuable data.

Our next area of analysis is the amalgamation of two transformative forces — the Internet of Things (IoT) and blockchain. Explore the synergies that arise when interconnected devices communicate securely through the decentralized architecture of the blockchain. Gaming takes center stage next, where we examine the symbiotic relationship between blockchain and the gaming industry. Through the lens of Non-Fungible Tokens (NFTs), witness how digital assets are transforming the gaming landscape, paving the way for a new era of ownership and creativity.

The narrative evolves further, where we scrutinize the integration of Regulatory Technology (RegTech) with blockchain. As technology meets governance, explore how this amalgamation streamlines compliance processes, offering a glimpse into the future of regulatory efficiency. In the concluding chapter of Part II, we reflect on the echoes of progress resonating through the cryptoverse. These chapters serve as waypoints in our expedition, unraveling the multifaceted layers of innovation, challenges, and the dance between technology and regulation.

Our journey through the cryptoverse is an evolving narrative — a testament to the dynamism and transformative potential of blockchain technology. As we move forward, the chapters ahead promise to unveil new horizons, introducing you to the ever-expanding frontier of possibilities within the cryptoverse. So, fasten your seatbelts, and let the exploration continue.

Chapter 11: Smart Contracts- Automated Business Logic

Introduction:

In the ever-evolving landscape of blockchain technology, smart contracts have emerged as one of the most revolutionary and transformative innovations. These self-executing contracts with the terms of the agreement directly written into code have reshaped the way we envision and execute agreements. In this chapter, we'll examine the intricacies of smart contracts, understanding their functionality, querying mechanisms, and the promising future they hold.

Smart contracts are self-executing contracts with the terms of the agreement directly written into code. They run on blockchain technology and automatically enforce, verify, or execute the terms of a contract when predefined conditions are met. This decentralized, tamper-resistant nature eliminates the need for intermediaries, ensuring transparency and trust in transactions.[42]

How Smart Contracts Function:

Smart Contracts: programs on the blockchain that run when specific conditions are met. (IBM, infra)

At their core, smart contracts are powered by blockchain networks. They are programmed to execute predefined actions when specific conditions are met.[43] These conditions are typically written in code that runs on the blockchain. The decentralized nature of blockchain ensures that

[42] Kukkuru, Manjuntha (2023). "Smart Contract: introducing a Transparent Way to Do Business." Infosys at https://www.infosys.com/insights/digital-future/smart-contracts.html.
[43] Frankenfield (2023), *supra*.

once the conditions are met, the contract is automatically executed without the need for a central authority.

Smart contracts utilize cryptographic keys to secure transactions and ensure the integrity of the contract. The immutability of blockchain technology guarantees that once a smart contract is deployed, its code cannot be altered or tampered with, providing a high level of security. Interacting with smart contracts involves querying their state or triggering specific functions. Users can interact with smart contracts through transactions, sending messages to the contract's address. These messages may contain specific instructions or queries.[44]

The Ethereum blockchain, a prominent platform for smart contracts, utilizes the Ethereum Virtual Machine (EVM) to execute smart contract code.[45] Developers and users can query the state of a smart contract by accessing its data stored on the blockchain. This transparency ensures that the current state of a contract is visible to all participants.[46]

Benefits of Smart Contracts:

Beyond their foundational functionality, smart contracts offer several benefits. One notable advantage is automation. Smart contracts automate the execution of contractual terms, reducing the need for intermediaries and streamlining processes. This not only enhances efficiency but also minimizes the risk of human error.

Transparency is another key benefit. The decentralized nature of blockchain ensures that the code and execution of smart contracts are visible to all relevant parties. This transparency

[44] IBM, *Infra*.
[45] Staff (2023). "Understanding Different ETH Contract Types." NFT Culture at https://www.nftculture.com/nft-news/understanding-different-eth-contract-types-721-vs-1155-vs-20/.
[46] Frankenfield (2023), *supra*.

fosters trust among participants, as they can independently verify the terms and outcomes of a contract.

Smart contracts facilitate trustless transactions. Trustless doesn't imply a lack of trust but rather a reliance on code and mathematics to execute agreements. Participants can engage in transactions without needing to trust a central authority, relying on the predetermined logic of the smart contract.

Challenges and Limitations:

While smart contracts offer tremendous potential, they are not without challenges and limitations. One significant challenge is the "oracle problem," where smart contracts rely on external information.[47] Oracles act as bridges between the blockchain and real-world data, introducing a potential point of failure or manipulation.

Another limitation is the irreversible nature of smart contracts. Once deployed, their code is immutable, meaning any errors or vulnerabilities cannot be easily rectified. This immutability necessitates thorough testing and auditing during the development phase to prevent potential issues.

Future Uses of Smart Contracts:

As blockchain technology continues to mature, the potential applications for smart contracts are expanding. Beyond their current use in areas like decentralized finance (DeFi) and token sales, smart contracts are poised to revolutionize various industries.

In supply chain management, smart contracts can automate and streamline processes, ensuring transparency and traceability. In healthcare, they could facilitate secure and transparent sharing of patient data. Real estate transactions, legal agreements, and

---

[47] Staff (2023). "The Blockchain Oracle Problem." Chainlink at https://chain.link/education-hub/oracle-problem.

even voting systems are areas where smart contracts can bring increased efficiency and trust.

Conclusion:

In conclusion, smart contracts represent a paradigm shift in how we conceptualize and execute agreements. Their self-executing nature, powered by blockchain, introduces a new era of transparency, efficiency, and trust in various domains. As we navigate this landscape of decentralized automation, the potential applications of smart contracts seem boundless, promising a future where agreements are not just written but encoded into the very fabric of our digital interactions.

Chapter 12: Decentralized Finance (DeFi)- Revolutionizing Traditional Finance

Introduction:

Decentralized Finance (DeFi) has emerged as a disruptive force in the financial landscape, challenging traditional notions of banking and finance. Operating on blockchain technology, DeFi leverages the principles of decentralization to offer a wide array of financial services. In this chapter, we'll explore the fundamentals of DeFi, its workings, and the various forms of governance that underpin this revolutionary financial ecosystem.

Decentralized Finance Unveiled:

> *Decentralized Finance: the technology that moves banking from a centralized systems towards a decentralized system to enable more ease of use, functionality, and inclusion.*

DeFi refers to a set of financial services and applications built on blockchain technology, primarily Ethereum. Unlike traditional finance, DeFi operates in a decentralized manner, eliminating the need for intermediaries like banks and enabling direct peer-to-peer transactions.[48] It aims to create an open and inclusive financial system that is accessible to anyone with an internet connection.

At the heart of DeFi is the use of smart contracts, self-executing contracts with the terms directly written into code. These contracts automate various financial processes, including lending, borrowing, trading, and yield farming. Smart contracts, residing on blockchain platforms like Ethereum, ensure

[48] Sharma, Rakesh (2023). "What is Decentralized finance (DeFI) and How does it work." Investopedia at https://www.investopedia.com/decentralized-finance-defi-5113835.

transparency, security, and efficiency in these financial transactions.

Forms of Governance in DeFi:

Decentralized finance relies on various governance mechanisms to make key decisions and updates to the protocol. Governance in DeFi can take several forms:

1. **Token-based Governance:** Many DeFi projects employ token-based governance, where holders of the project's native tokens can participate in decision-making. The more tokens one holds, the more influence they have in shaping the protocol's future. This model aligns the interests of token holders with the success of the platform.
2. **Community Governance:** Some DeFi projects opt for a more community-driven approach, where decisions are made through discussion forums, signaling mechanisms, and community proposals. This approach fosters a sense of community ownership and participation.
3. **Decentralized Autonomous Organizations (DAOs):** DAOs are entities represented by rules encoded as a computer program. They are often used in DeFi for collective decision-making. Token holders in a DAO can propose, discuss, and vote on changes to the protocol. DAOs aim to distribute power among community members.

The Future Landscape of DeFi:

As DeFi continues to evolve, its impact on the financial industry becomes increasingly significant. The transparency, accessibility, and efficiency offered by decentralized finance have the potential to reshape the traditional banking landscape. DeFi protocols are exploring interoperability, scalability, and user experience enhancements to attract a broader audience.

Conclusion:

In conclusion, decentralized finance represents a groundbreaking shift in the financial paradigm. By leveraging blockchain technology and innovative governance models, DeFi opens up financial services to a global audience, providing an alternative to traditional banking systems. As the ecosystem matures, the principles of transparency, inclusivity, and community-driven decision-making will likely play a pivotal role in shaping the future of decentralized finance.

Chapter 13: Tokenization- Unlocking and Digitizing Asset Value

Introduction:

Tokenization has emerged as a transformative force within the cryptoverse and blockchain, revolutionizing the way we represent and transfer value. This chapter explores the concept, methodology, and functionality of tokenization, shedding light on its diverse applications and the impact it has on various industries. As we navigate the complexities of tokenization, we unravel its potential to reshape traditional systems and foster innovation.

At its core, tokenization involves the representation of real-world assets or rights on a blockchain in the form of digital tokens. These tokens, often standardized through smart contracts, enable the fractional ownership or representation of assets that range from real estate and art to financial instruments and intellectual property. The process of tokenization imbues these assets with newfound liquidity, accessibility, and programmability.

Methodology of Tokenization:

The methodology of tokenization encompasses several key steps. Firstly, the asset to be tokenized undergoes legal and technical assessments to ensure its suitability for representation on a blockchain. Once deemed viable, the asset is divided into tradable and transferable digital tokens, each representing a fractional ownership stake.[49] Smart contracts govern the rules and conditions of these tokens, automating processes such as dividend distribution or voting rights.

---

[49] Awan, Abid All (2023). "What is Tokenization?" Data Camp at https://www.datacamp.com/blog/what-is-tokenization.

Blockchain platforms, especially those supporting smart contracts like Ethereum, have become instrumental in facilitating tokenization. These platforms provide the infrastructure and tools needed to deploy and manage tokenized assets securely. Additionally, regulatory compliance plays a pivotal role in shaping the methodology of tokenization, with adherence to legal frameworks ensuring legitimacy and investor protection.[50]

Functionality of Tokenization:

Tokenization introduces a range of functionalities that redefine the s of ownership, investment, and transferability. One of its primary functions is fractional ownership, allowing individuals to own a percentage of high-value assets that would otherwise be financially prohibitive.[51] This democratization of ownership expands investment opportunities to a broader audience.

Liquidity is another key functionality of tokenization. Traditional assets, such as real estate, often lack liquidity due to their illiquid nature. Tokenization transforms these assets into tradable tokens, enhancing their marketability and enabling investors to buy or sell their holdings more readily.

Liquidity: Availability of the asset or its underlying value to be transferred to another.

Programmability is a distinctive feature of tokenization.[52] Smart contracts embedded within tokens can automate various processes, from dividend distribution to the execution of complex financial agreements. This programmability not only

---

[50] *Id.*

[51] Lutkevich, Ben (2023). "Tokenization." Tech Target at https://www.techtarget.com/searchsecurity/definition/tokenization.

[52] *Id.*

enhances efficiency but also reduces the need for intermediaries in certain transactions.

Applications of Tokenization:

Tokenization finds applications across diverse industries, bringing about innovative solutions and disrupting traditional models. In real estate, tokenization allows for the fractional ownership of properties, making real estate investment more accessible. Art and collectibles can be tokenized, enabling investors to own a share of valuable artworks.

Financial markets benefit from tokenization through the representation of stocks, bonds, and other instruments as digital tokens. This enhances market efficiency, reduces settlement times, and introduces new avenues for global investment.[53]

The tokenization of intellectual property, such as patents or copyrights, transforms intangible assets into tradable tokens, allowing creators to monetize their work more efficiently. Tokenization also extends to the gaming industry, where in-game assets can be tokenized and traded on blockchain-based marketplaces.

Challenges and Considerations:

Despite its transformative potential, tokenization is not without challenges. Regulatory compliance remains a significant consideration, with legal frameworks varying globally. Ensuring adherence to these regulations is crucial for the legitimacy and widespread acceptance of tokenized assets.

Interoperability between different blockchain networks poses challenges, as assets tokenized on one platform may face limitations in terms of transferability and liquidity on others. Security concerns, including the vulnerability of smart contracts

---

[53] *Id.*

to exploitation, underscore the importance of robust cybersecurity measures.

Conclusion:

In conclusion, tokenization stands as a force reshaping the landscape of ownership, investment, and transferability. Its methodology, driven by smart contracts and blockchain technology, facilitates the fractional representation of diverse assets. The functionalities of tokenization, including fractional ownership, enhanced liquidity, and programmability, unlock new possibilities across industries.

As tokenization continues to evolve, its applications diversify, offering innovative solutions to age-old challenges. While challenges such as regulatory compliance and interoperability persist, the potential benefits of tokenization in democratizing access to assets and fostering global investment underscore its significance in the ongoing transformation of traditional systems.

Chapter 14: Supply Chain Integration-Transparency and Traceability

Introduction:

Supply chain integration within the realms of blockchain and tokenomics has emerged as a driver of efficiency, transparency, and trust in modern business ecosystems. This chapter looks into the profound significance of integrating supply chains with blockchain technology and tokenomics, elucidating the transformative impact on various industries. Navigating the intersections of these technologies, we uncover the potential to revolutionize how we perceive, manage, and optimize supply chains.

The Imperative of Supply Chain Integration:

Supply chains are networks involving the production, distribution, and consumption of goods and services. Traditionally, these chains have been plagued by inefficiencies, lack of transparency, and susceptibility to fraud.[54] The integration of blockchain technology and tokenomics addresses these challenges by providing a decentralized, transparent, and traceable framework for managing the flow of goods and information.

Blockchain's Role in Supply Chain Integration:

Blockchain serves as the backbone of supply chain integration, offering a decentralized ledger that records and verifies transactions across a network of nodes. This distributed nature ensures that all participants in the supply chain have real-time

---

[54] Staff (2023). "Role of Web3 in Supply Chain Management and Logistics Management." Maticz at https://maticz.com/web3-in-supply-chain-logistics.

access to a single version of the truth.[55] Every transaction, from manufacturing to distribution to retail, is recorded in a secure and tamper-resistant manner.

The transparency provided by blockchain fosters trust among stakeholders. Each participant can trace the journey of a product from its origin to the end consumer. This visibility is particularly crucial in sectors like food and pharmaceuticals, where consumers demand accurate information about the provenance and authenticity of products.[56]

Blockchain's decentralized consensus mechanism ensures that data cannot be altered retroactively. This immutability enhances the integrity of supply chain records, mitigating the risk of fraud and ensuring the accuracy of information.[57] Smart contracts, self-executing contracts with coded terms, automate various processes in the supply chain, reducing the need for intermediaries and enhancing efficiency.

Tokenomics and Supply Chain Optimization:

Tokenomics, the economic model governing the creation, distribution, and management of tokens, complements blockchain in optimizing supply chains.[58] Tokens can represent ownership stakes, incentive mechanisms, or access rights within the supply chain network. By leveraging tokenomics, organizations can introduce innovative incentives, streamline

---

[55] Staff (2023). "Role of Web3 in Supply Chain Management." Solu Lab at https://www.solulab.com/web3-in-supply-chain-and-logistics/.

[56] *Id.*

[57] Staff (2024). "Supply Chain Unleashing the power of Web3 Technology." Tech.blog at https://niaai.tech.blog/2024/01/10/supply-chain-unleashing-the-power-of-web3-technology/.

[58] Notomoro (2023). "A Deep Dive into the Future of Web3 in Supply Chain Management." WebiSoft at https://webisoft.com/articles/web3-in-supply-chain-and-logistics/.

transactions, and create a more collaborative and efficient ecosystem.

Incentive mechanisms, facilitated through tokens, encourage stakeholders to actively participate in the supply chain. For instance, suppliers delivering products ahead of schedule or maintaining high-quality standards can be rewarded with tokens. This not only motivates better performance but also creates an ecosystem where participants are aligned with the overall success of the supply chain.[59]

Tokenized assets within the supply chain enable fractional ownership and trading. This introduces liquidity to traditionally illiquid assets, allowing stakeholders to monetize their holdings or diversify their portfolios.[60] Tokenomics also facilitates the creation of tokenized supply chain finance, where tokens represent financial instruments, providing liquidity to participants at various stages of the supply chain.

Traceability and Accountability:

One of the primary advantages of supply chain integration with blockchain and tokenomics is the enhanced traceability and accountability it affords. Every step of the supply chain process, from raw material sourcing to manufacturing to distribution, is recorded in an immutable ledger.[61] This transparency ensures that any discrepancies or issues can be quickly identified and addressed.

The accountability introduced by blockchain and tokenomics extends beyond the organization to the entire supply chain network. Smart contracts can automate compliance with regulatory standards, ensuring that all participants adhere to predefined rules and regulations. This not only reduces the risk

---

[59] *Id.*
[60] Solu Labs (2023), *supra*.
[61] *Id.*

of regulatory violations but also fosters a culture of accountability among stakeholders.

Challenges and Considerations:

While the integration of blockchain and tokenomics in supply chains holds immense potential, challenges persist. Interoperability between different blockchain networks remains a consideration, as participants may use different platforms. Standardization efforts are underway to address this challenge and ensure seamless collaboration.

Integration also requires significant investments in technology infrastructure and education. Organizations need to train their workforce to adapt to the new technological landscape and embrace the benefits of blockchain and tokenomics fully. Additionally, regulatory frameworks around tokenized assets need further development to provide legal clarity and support widespread adoption.

Conclusion:

In conclusion, the integration of supply chains with blockchain technology and tokenomics represents a shift in how we conceptualize and optimize these networks. Blockchain's decentralized ledger ensures transparency, immutability, and traceability, addressing historical inefficiencies and fostering trust among stakeholders.

Tokenomics introduces innovative economic models, incentivizing participation, optimizing transactions, and creating a more and collaborative supply chain ecosystem. The combined impact of blockchain and tokenomics extends beyond individual organizations, reshaping entire industries by introducing unprecedented levels of transparency, accountability, and efficiency to global supply chains. As organizations continue to navigate this journey, the potential benefits of integrated supply chains will likely redefine the future of commerce and trade.

Chapter 15: Identity Management- Securing Digital Identities

Introduction:

Identity management plays an important role in the landscape of blockchain, establishing its nature in ensuring security, adhering to regulatory frameworks, and validating transactions. This chapter explores the multifaceted significance of identity management within the blockchain realm, delving into the intricate balance between security measures, Know Your Customer (KYC) protocols, and the validation of transactions. As we navigate the nuances of identity within the blockchain space, we unravel the transformative potential of decentralized identity solutions.

Identity Security in the Blockchain Era:

KYC: Know your Client/Customer, an important role in preventing criminal activity.

In the conventional digital landscape, identity theft, fraud, and unauthorized access have been persistent challenges. Blockchain, with its decentralized and cryptographic foundations, provides a robust solution to enhance identity security.[62] The use of cryptographic keys ensures that identities are protected by advanced encryption, making it exceedingly difficult for malicious actors to compromise personal information.[63]

---

[62] Staff (2023). "The Impact of Blockchain on Digital Identity Management: A New Era of Trust." Syednaadeali at https://medium.com/@syednaadeali512/the-impact-of-blockchain-on-digital-identity-management-a-new-era-of-trust-afc5296d5cd2.

[63] Id.

Blockchain's distributed nature further fortifies identity security. Traditional centralized systems are vulnerable to single points of failure, making them attractive targets for cyber-attacks. In contrast, a decentralized identity management system disperses identity data across a network of nodes, reducing the risk of unauthorized access and enhancing overall security.

The immutability of blockchain ensures that once identity information is recorded, it cannot be altered or tampered with retroactively. This tamper-resistant quality enhances the integrity of identity records, providing a secure and unforgeable foundation for user identification.

KYC Protocols and Compliance:

Know Your Customer (KYC) protocols have long been a cornerstone of regulatory compliance, particularly in sectors like finance and healthcare.[64] Blockchain facilitates the implementation of KYC processes in a decentralized manner, ensuring compliance while preserving user privacy.

Decentralized identity solutions leverage blockchain's capabilities to enable self-sovereign identity, empowering individuals to control their personal information. Users can selectively disclose specific details for KYC purposes without compromising the entirety of their identity.[65] This selective disclosure not only aligns with privacy principles but also streamlines the KYC process for both individuals and organizations.

Smart contracts, self-executing contracts with coded terms, can automate KYC processes by verifying user information against

---

[64] Chen, James (2024). "Know your client KYC: What it means, Compliance Requirements." Investopedia at https://www.investopedia.com/terms/k/knowyourclient.asp.

[65] Staff (2023). "KYC: 3 Steps to Achieving Know Your Customer Compliance." Truilioo at https://www.trulioo.com/blog/kyc.

predefined criteria. This reduces manual intervention, accelerates the onboarding process, and ensures compliance with regulatory requirements. Additionally, the transparency of blockchain allows regulators to audit and verify KYC processes, fostering a more accountable ecosystem.[66]

Transaction Validation and Identity:

In the context of blockchain, identity management is closely intertwined with the validation of transactions. Traditional centralized systems rely on intermediaries to verify and authenticate transactions, introducing delays and costs. Blockchain's decentralized nature enables the validation of transactions through consensus mechanisms, eliminating the need for intermediaries.

Public blockchains utilize consensus mechanisms like Proof of Work (PoW) or Proof of Stake (PoS) to validate transactions. Participants in the network, known as nodes, reach consensus on the validity of transactions through cryptographic algorithms.[67] Identity management plays a crucial role in ensuring that only authorized nodes participate in the validation process.

Private blockchains, often employed in enterprise settings, implement permissioned systems where only authorized entities can participate as validating nodes. Identity verification

---

[66] B. Harry (2023). "Digital Identity and Blockchain: A New Era of Identity Verification" Linked In at https://www.linkedin.com/pulse/digital-identity-blockchain-new-era-verification-harry-b--exepf/?trk=article-ssr-frontend-pulse_more-articles_related-content-card.

[67] Staff (2023). "What is the difference between validation and verification." Experian at https://www.experian.co.uk/blogs/latest-thinking/fraud-prevention/what-is-the-difference-between-validation-verification-and-authentication/.

becomes integral to maintaining the integrity of the network and validating transactions securely.

Challenges and Considerations:

While identity management on the blockchain offers substantial benefits, challenges and considerations persist. Scalability remains a concern, especially on public blockchains with a large user base. Balancing the demand for transaction throughput with robust identity verification processes requires ongoing innovation and optimization.

Interoperability between different blockchain networks and standardization of identity management protocols are critical considerations. Efforts to establish common standards for decentralized identity solutions are underway, aiming to create a cohesive and interoperable ecosystem.

The evolving regulatory landscape surrounding digital identity also poses challenges. Striking the right balance between compliance and user privacy necessitates ongoing dialogue between blockchain innovators, regulators, and end-users.

Conclusion:

In conclusion, identity management within the blockchain landscape transcends traditional notions, offering a paradigm shift in how we secure, validate, and verify user information. Blockchain's decentralized and cryptographic foundations fortify identity security, making it resistant to unauthorized access and tampering.

The integration of KYC protocols into decentralized identity solutions aligns with regulatory requirements while empowering individuals with control over their personal information. Smart contracts automate KYC processes, enhancing efficiency and accountability in user onboarding.

Transaction validation, a core function of blockchain, relies on robust identity management to ensure the integrity of decentralized networks. Whether through public blockchains with consensus mechanisms or private blockchains with permissioned systems, identity verification plays a pivotal role in maintaining the trust and security of transactions.

As identity management continues to evolve within the blockchain space, addressing scalability, interoperability, and regulatory considerations will be crucial. The ongoing journey towards a decentralized, secure, and user-centric identity paradigm promises transformative outcomes, reshaping how we navigate the digital landscape with trust and confidence.

Chapter 16: Cross Border Payments-Protecting Sensitive Information

Introduction:

Cross-border payments stand at the intersection of commerce and connectivity, a vital conduit for global economic activities. This overview serves as a prelude to our detailed exploration in Part V, where we dissect the intricacies of cross-border transactions within the cryptoverse. Traditional cross-border systems often grapple with inefficiencies, delays, and cost complexities, prompting the exploration of blockchain-based solutions. The decentralized and borderless nature of cryptocurrencies and blockchain technology promises to revolutionize cross-border payments, ushering in an era of efficiency and financial inclusion.

The Landscape of Cross-Border Payments:

Cross-border transactions, traversing the labyrinth of traditional financial systems, face challenges ranging from prolonged processing times to exorbitant fees and currency conversions. The cryptoverse emerges as a potential disruptor, offering solutions that leverage blockchain's speed, transparency, and cost-effectiveness.[68] As we journey through the cryptoverse's landscape, we'll unravel the potential transformations and address the challenges inherent in cross-border payments.

Blockchain's Promise for Cross-Border Payments:

The decentralized nature of blockchain eliminates the need for intermediaries, reducing processing times and associated costs. Cryptocurrencies, operating on blockchain networks, enable

---

[68] Lakhman, Ira (2023). "Overcoming Cross-Border Payment Hurdles: How Web3 is Transforming Payments." Business Day at https://businessday.ng/bd-weekender/article/overcoming-cross-border-payment-hurdles-how-web3-is-transforming-payments/.

swift and transparent cross-border transactions, challenging the limitations of traditional banking systems. Smart contracts, programmable agreements executed on blockchain platforms, add an additional layer of efficiency to cross-border payments.[69]

Challenges and Considerations:

While the cryptoverse offers promises, it also poses challenges. Regulatory uncertainties, volatility in cryptocurrency values, and concerns about security and privacy demand careful consideration. Navigating these challenges requires a nuanced understanding of the evolving landscape and a proactive approach to address potential pitfalls.

The Intersection of Cryptocurrencies and Cross-Border Transactions:

In Part V, we will dissect the symbiotic relationship between cryptocurrencies and cross-border payments. Examining real-world use cases and innovations, we aim to provide a comprehensive understanding of how blockchain technology is reshaping the s of international transactions. From remittances to global trade, the cryptoverse offers a spectrum of possibilities that redefine the way we perceive and conduct cross-border payments.

Conclusion:

This overview lays the foundation for our in-depth exploration in Part V, where the spotlight turns towards the transformative potential and challenges embedded in cross-border payments within the cryptoverse. As we embark on this journey, anticipate insights that illuminate the evolving landscape of international

---

[69] Singh, Shailey (2023). "What are cross-border payments, and how do they work?" Coin Telegraph at https://cointelegraph.com/explained/what-are-cross-border-payments-and-how-do-they-work.

transactions, unveiling the cryptoverse's role in shaping the future of cross-border payments.

Chapter 17: Data Privacy and Security-Protecting Sensitive Data

Introduction:

The cryptoverse, a realm built on the pillars of decentralization and transparency, grapples with a delicate balancing act between data security and privacy. As blockchain and cryptocurrency technologies redefine how we interact with data, understanding this equilibrium becomes paramount. In this exploration, we consider the interplay, dissecting the mechanisms that safeguard sensitive information while upholding the principles of privacy within the cryptoverse.

The Foundations of Data Security:

Data security, a fundamental tenet of the cryptoverse, relies on cryptographic principles embedded in blockchain networks. Through advanced encryption algorithms, blockchain ensures the integrity and confidentiality of data. Immutable ledgers, a hallmark of blockchain, provide an additional layer of security, rendering data tamper-resistant.[70] However, the challenge lies in maintaining this robust security infrastructure without compromising user privacy.

The transparency inherent in blockchain's distributed ledger technology presents a paradox. While transactions are visible to all participants, preserving the anonymity and privacy of users becomes a nuanced challenge. Cryptocurrencies like Bitcoin operate pseudonymously, associating transactions with cryptographic addresses rather than personal identities. Yet, achieving true privacy within a transparent system remains an ongoing pursuit.

In response to the privacy conundrum, privacy-focused cryptocurrencies like Monero and Zcash have emerged. These

---

[70] Howell, *infra*.

coins implement sophisticated cryptographic techniques, such as ring signatures and zero-knowledge proofs, to obfuscate transaction details.[71] Confidential transactions, a privacy-enhancing feature, secure transaction amounts without revealing them on the public ledger, fostering a more private financial ecosystem.

Challenges and Concerns:

While privacy-centric solutions offer advancements, challenges persist. Regulatory scrutiny, particularly around privacy coins, poses a threat to their widespread adoption. Striking a balance between complying with regulations and preserving user privacy becomes a delicate dance within the cryptoverse.[72] Additionally, concerns related to illicit activities and money laundering necessitate nuanced solutions that address security without compromising the essence of privacy.

Empowering users with control over their data lies at the heart of the cryptoverse's vision for privacy. Self-sovereign identity solutions leverage blockchain to give individuals ownership and control over their personal information.[73] By decentralizing identity management, users can selectively disclose information, enhancing privacy without sacrificing security.

As the cryptoverse matures, privacy becomes a focal point of innovation. Solutions like privacy-focused smart contracts and decentralized finance (DeFi) platforms are redefining how data is handled within blockchain ecosystems.[74] Striking a balance

---

[71] Staff (2023). "The Future of Privacy in a Web3 World." Shardeum at https://shardeum.org/blog/privacy-in-web3/.
[72] Staff (2023). "Understanding Web3 and its Security Implications." CloudFlare at https://www.cloudflare.com/the-net/web3-security/.
[73] Vaughan, Victoria (2023). "Top Web3 Privacy Challenges and How to Avoid them." Be In Crypto at https://beincrypto.com/learn/web3-privacy-challenges/.
[74] *Id.*

between compliance, security, and privacy is a process, shaped by technological advancements and evolving regulatory landscapes.

Conclusion:

In navigating the cryptoverse's landscape, the delicate relationship between data security and privacy unfolds. The inherent tension between transparency and confidentiality propels the cryptoverse towards innovative solutions. As we conclude this exploration, the cryptoverse stands at a crossroads, navigating the path towards a future where the security-privacy equilibrium remains a cornerstone, fostering a trusted and resilient digital ecosystem.

Chapter 18: Internet of Things (IoT) Integration- Smart Contracts for Devices

Introduction:

The symbiotic integration of the Internet of Things (IoT) and blockchain within the cryptoverse marks a pivotal convergence, promising to redefine how devices communicate, transact, and secure data. In this exploration, we unravel the mystery that emerges when these transformative technologies intertwine, ushering in an era where the cryptoverse extends its reach into the physical realm through the vast network of interconnected devices.

IoT and Blockchain- A Symbiotic Nexus:

At its core, the IoT is an expansive ecosystem of interconnected devices, from everyday objects to industrial machinery, designed to collect and exchange data. Blockchain, a decentralized ledger technology, introduces a layer of trust and security.[75] As these technologies converge, a symbiotic nexus forms, addressing crucial challenges in data integrity, security, and the establishment of a decentralized and tamper-resistant record.

Traditional centralized systems in the IoT pose vulnerabilities, as a single point of failure can compromise the entire network. Blockchain's decentralized architecture eliminates this vulnerability. Each IoT device transaction becomes a block in the chain, distributed across nodes.[76] This not only ensures data integrity but also mitigates the risk of unauthorized tampering, enhancing the overall robustness of the IoT network.

---

[75] Staff (2024). “Build Trust in your IoT data with Blockchain.” IBM at https://www.ibm.com/topics/blockchain-iot.

[76] Deloitte at https://www2.deloitte.com/ch/en/pages/innovation/articles/blockchain-accelerate-iot-adoption.html.

Security concerns have been a persistent challenge in the IoT landscape. Blockchain introduces cryptographic principles that fortify the security of device communication. Smart contracts, self-executing code on the blockchain, enable secure and automated interactions between IoT devices. This cryptographic layer ensures that only authorized devices can participate in transactions, fostering a secure environment.[77]

Blockchain provides a solution to the challenge of managing and verifying device identities. Each IoT device is assigned a unique cryptographic identity stored on the blockchain. This immutable identity ensures authenticity and enables transparent verification. As devices autonomously transact on the blockchain, their identities are cryptographically secured, preventing identity spoofing and unauthorized access.

Efficient Supply Chain Management:

The integration of IoT and blockchain revolutionizes supply chain management. IoT sensors can continuously monitor and record the journey of goods from manufacturing to delivery. This real-time data is immutably recorded on the blockchain, providing stakeholders with transparent and verifiable information.[78] From traceability of product origins to real-time updates on transportation conditions, the cryptoverse enhances supply chain efficiency.

In the realm of energy, the convergence of IoT and blockchain fosters decentralized energy management. IoT devices, such as smart meters and sensors, can autonomously transact energy data on the blockchain. Smart contracts facilitate peer-to-peer energy transactions, allowing devices to buy and sell energy in a

---

[77] IBM (2024), *supra*.

[78] Le, Vy (2023). "IoT in supply chain management: How IoT Empowers Supply Chain at Every Step." Orient at https://www.orientsoftware.com/blog/iot-in-supply-chain/.

decentralized manner. This disintermediation of energy transactions promotes efficiency and reduces costs.

Despite the transformative potential, challenges exist. Scalability, interoperability, and standardization across IoT devices are critical considerations.[79] Overcoming these hurdles requires collaborative efforts and industry-wide initiatives to establish common protocols that facilitate seamless integration.

As the integration of IoT and blockchain within the cryptoverse evolves, the future landscape holds promise.[80] Smart cities, autonomous vehicles, and connected industries represent domains where this symbiotic convergence will reshape how data is generated, transmitted, and secured. The cryptoverse's influence extends beyond the digital, weaving into the fabric of our physical surroundings through the interconnected system of IoT devices.

Conclusion:

In concluding this exploration, the integration of IoT into the cryptoverse emerges as a transformative force, augmenting the capabilities of both technologies. From decentralized data integrity to secure device communication, the symbiotic nexus of IoT and blockchain paves the way for a future where the physical and digital realms seamlessly intertwine. The cryptoverse's expansion into the vast landscape of interconnected devices signifies not only technological evolution but also the convergence of two powerful forces shaping our connected future.

---

[79] Gaur, Vishal (2021). "Brining Blockchain, IoT, and Analytics to Supply Chains." Harvard Business Review at https://hbr.org/2021/12/bringing-blockchain-iot-and-analytics-to-supply-chains.

[80] *Id.*

Chapter 19: Gaming and Non-Fungible Tokens (NFTs)- Monetizing Virtual Assets

Introduction:

The intersection of blockchain technology and the gaming industry has given rise to a revolutionary paradigm, redefining how gamers engage with virtual worlds and digital assets. In this exploration, we look into the relationship between gaming ecosystems and blockchain, examining how non-fungible tokens (NFTs) have become a pivotal force, unlocking new dimensions of ownership, authenticity, and value within the cryptoverse.

Gaming and Blockchain: A Symbiotic Evolution:

The gaming industry, with its vast user base and virtual economies, has found a natural ally in blockchain technology. Blockchain's decentralized and transparent nature addresses longstanding challenges in gaming, introducing novel concepts that redefine player experiences. Decentralized ownership, verifiable scarcity, and provable authenticity are at the forefront of this symbiotic evolution.

Traditional gaming models often limit players' ownership of in-game assets. Blockchain disrupts this paradigm by introducing decentralized ownership. NFTs, representing unique digital assets on the blockchain, empower gamers with true ownership.[81] Whether it's rare in-game items, characters, or skins, blockchain ensures that players have verifiable ownership and control over their virtual possessions.

The tokenization of in-game assets through NFTs opens new avenues for value creation. Gamers can tokenize their rare items or achievements, creating a digital representation on the

---

[81] Leroy, Kaj (2023). "The Future of NFTs: Gaming and IP." Forbes at https://www.forbes.com/sites/forbesbusinesscouncil/2023/02/24/the-future-of-nfts-gaming-and-ip/?sh=4372cee24670.

blockchain. These tokens can be bought, sold, or traded across gaming ecosystems, fostering a secondary market for virtual assets. This introduces a new dimension to the gaming experience, where virtual possessions hold tangible value.

Provably Scarce and Authentic Items:

Scarcity and authenticity are intrinsic to the appeal of rare in-game items. Blockchain ensures the provable scarcity of NFTs, as each token is unique and its scarcity is verifiable on the blockchain. This brings a level of authenticity to virtual items, eliminating concerns of duplication or counterfeit in the gaming space.[82]

Blockchain-driven gaming ecosystems empower players to participate in player-driven economies. Through decentralized marketplaces, gamers can trade in-game assets directly with one another, creating a peer-to-peer economy. This shift towards player-driven economies not only enhances the gaming experience but also introduces economic incentives for skilled players and content creators.

While the broader concept of NFTs is reserved for detailed discussion later in the book, their profound impact on gaming deserves a deeper dive. NFTs enable the tokenization of unique in-game items, skins, characters, and even entire game environments. This unique representation on the blockchain brings an unprecedented level of rarity, ownership, and tradability to virtual assets.

Despite the transformative potential, challenges exist. Scalability, interoperability across gaming platforms, and environmental concerns related to certain blockchain networks

---

[82] *Id.*

are areas that require ongoing attention.[83] However, these challenges present opportunities for collaborative innovation within the gaming and blockchain space.

As blockchain and gaming continue to converge, the future promises a landscape where virtual worlds become interconnected ecosystems of value and ownership. The integration of blockchain and NFTs into gaming extends beyond traditional gaming experiences, creating a metaverse where players are not just participants but active contributors to the evolving narrative of virtual realms.

Conclusion:

In conclusion, the fusion of gaming and blockchain, fueled by the introduction of NFTs, signifies a transformative era for the gaming industry. The decentralization of ownership, tokenization of virtual assets, and the establishment of player-driven economies lay the groundwork for a future where the lines between the virtual and physical blur. As blockchain technology continues to shape the gaming landscape, players are no longer mere participants but pioneers in an evolving cryptoverse where digital ownership takes center stage.

[83] Clere, Alex (2023). "Could NFTs be the gaming industries next big revenue Stream?" Fintech at https://fintechmagazine.com/articles/could-nfts-be-the-gaming-industrys-next-big-revenue-stream.

Chapter 20: Regulatory Technology (RegTech)- Navigating Compliance

Introduction:

The integration of regulatory technology (RegTech) and blockchain has emerged as a pivotal development, reshaping the landscape of compliance, transparency, and regulatory adherence within the cryptoverse. In this chapter, we navigate the intersection of RegTech and blockchain technology, exploring how decentralized systems can enhance regulatory processes, mitigate risks, and foster a compliant environment.

> *RegTech: the application of new technology to manage business compliance.*

RegTech, with its focus on leveraging technology to streamline regulatory compliance, finds a natural ally in the decentralized and transparent attributes of blockchain.[84] The synergistic alliance between RegTech and blockchain introduces novel methodologies for regulatory oversight, data management, and compliance automation.

Enhancing Regulatory Compliance with Blockchain:

Blockchain's inherent features, including immutability, transparency, and cryptographic security, contribute to the enhancement of regulatory compliance. Smart contracts, self-executing contracts with predefined conditions, can automate

---

[84] Staff (2023). "Embracing the Future: How AI, Regtech, and Blockchain are Revolutionizing the Fintech Landscape." OMA Emirates Group at https://www.linkedin.com/pulse/embracing-future-how-ai-regtech-blockchain-revolutionizing/.

regulatory processes, ensuring adherence to predefined rules without the need for centralized oversight.[85]

Immutable Record-Keeping and Auditing:

One of the key challenges in regulatory compliance is maintaining accurate and tamper-proof records. Blockchain's decentralized ledger provides an immutable record-keeping system, where each transaction or regulatory event is timestamped, cryptographically secured, and linked to previous entries.[86] This ensures a transparent and auditable trail of regulatory activities.

Identity Management and KYC on the Blockchain:

Regulatory requirements often involve stringent identity verification and Know Your Customer (KYC) procedures.[87] Blockchain facilitates secure and decentralized identity management, allowing individuals to control and share their identity information with regulatory bodies as needed. This not only streamlines KYC processes but also enhances data security and privacy.

Smart Contracts for Regulatory Reporting:

Smart contracts can automate regulatory reporting by executing predefined actions when specific reporting conditions are met. This reduces the manual effort and potential for errors in the reporting process.[88] Additionally, the transparent nature of

---

[85] Deloitte at https://www2.deloitte.com/mt/en/pages/risk/articles/mt-blockchain-in-compliance.html.

[86] Staff (2023). "The Importance of an Immutable Audit Trail in Distributed Ledgers." Faster Capital at https://fastercapital.com/content/The-Importance-of-an-Immutable-Audit-Trail-in-Distributed-Ledgers.html.

[87] Staff (2023). "Knowing Web3 and Blockchain Technology's KYC/AML concepts." Cyber Savvy at https://www.linkedin.com/pulse/kycaml-web3-blockchain-technology-laljan-basha-shaik/.

[88] Alchemy at https://www.alchemy.com/smart-contracts.

blockchain ensures that regulators can access real-time data without relying on intermediaries.

Tokenization of Compliance Instruments:

The tokenization of compliance instruments on the blockchain introduces a new dimension to regulatory processes. Compliance-related assets, such as licenses, permits, or certificates, can be represented as tokens on the blockchain. This facilitates easy tracking, transfer, and validation of compliance instruments, reducing bureaucratic inefficiencies.

While the integration of RegTech and blockchain offers significant benefits, challenges exist. Interoperability with existing systems, regulatory frameworks adapting to decentralized technologies, and concerns related to data privacy are aspects that require careful consideration. Collaboration between regulators, industry stakeholders, and technology innovators is essential to address these challenges. The trajectory of RegTech in the cryptoverse points towards a future where regulatory processes are seamlessly integrated with blockchain technology. As advancements continue, regulatory compliance can become more efficient, transparent, and adaptable to the nature of the blockchain industry.

Conclusion:

In conclusion, the convergence of RegTech and blockchain represents a shift in how regulatory compliance is approached within the cryptoverse. The marriage of decentralized technologies and regulatory processes brings efficiency, transparency, and security to compliance frameworks. As the industry evolves, the collaboration between RegTech and blockchain is poised to redefine the standards of regulatory adherence, paving the way for a compliant and innovative cryptoverse.

Conclusion Part II

As we conclude the exploration of Part II, delving into the intricacies of blockchain technology and its varied applications, a clear narrative emerges — the cryptoverse is not merely a digital frontier; it is a transformative force reshaping the foundations of technology, finance, and regulation.

In these chapters, we navigated the fundamental concepts and innovations that form the backbone of the cryptoverse. Smart contracts, with their self-executing capabilities, have redefined how agreements are conceptualized and executed. The decentralized finance (DeFi) ecosystem has unlocked new possibilities for financial services, offering a glimpse into a future where traditional intermediaries may become obsolete.

Tokenization emerged as a powerful tool, transcending industries like supply chain management and identity verification. It promises to reshape how we perceive and handle assets, adding layers of efficiency and transparency. Next, we glimpsed the complexities of cross-border payments, setting the stage for the challenges and opportunities that will unfold in Part V. As we examined the tightrope walk between data security and privacy, the critical role of identity management on the blockchain became evident.

In the domain of the Internet of Things (IoT) and blockchain integration, we unveiled a future where interconnected devices communicate seamlessly, underpinned by the trust and security of decentralized ledgers. Gaming and NFTs illustrated the intersection of blockchain technology with the vibrant world of digital assets and entertainment. The gamification of blockchain has opened avenues for innovation and user engagement. The marriage of RegTech and blockchain showcased the potential for technology to streamline regulatory compliance, reinforcing the narrative that the cryptoverse is not divorced from governance but rather aims for a harmonious integration.

As we reflect on these chapters, it becomes apparent that the cryptoverse is not a solitary entity; it is an ecosystem of interconnected technologies, each playing a distinct role in reshaping our digital landscape. The journey through Part II is just a glimpse into this ever-evolving terrain, and the forthcoming sections will unravel more layers of this ecosystem.

In the chapters to come, we will embark on new expeditions, exploring the challenges, innovations, and the evolving narrative of the cryptoverse. The cryptic codes of blockchain and decentralized technologies are gradually unveiling a future where the rules of engagement are being rewritten, and every participant has a role in this unfolding saga.

Part III: International Expeditions: Penetrating Foreign Markets with Blockchain

Introduction to Part III

In the labyrinth of global business, Chapters 21 to 30 unfold as a compass guiding us through the uncharted territories where blockchain technology intersects with international trade. This segment transcends the conventional boundaries of commerce, exploring the transformative impact of blockchain on global payment systems, the facilitation of international trade, supply chain s, cross-border contracts, regulatory compliance, tokenization, identity management, multi-lingual smart contracts, and the strategic nuances of international partnerships.

As we embark on this exploration, Chapter 21 serves as the genesis, emphasizing the pivotal role of global payment systems in the orchestration of international business. The narrative unravels how blockchain, with its decentralized architecture, reshapes the landscape of global transactions, promising efficiency, transparency, and security. While the curtain is raised on programs like SWIFT and Ripple/XRP, subsequent chapters pledge a more in-depth analysis. Chapter 22 ushers us into the realm where cryptocurrency becomes the bulwark for international trade. Beyond its role in cross-border transactions, we look into the multifaceted ways in which cryptocurrencies redefine the s of global commerce. This chapter encapsulates the essence of a financial revolution that transcends borders, fostering efficiency and trust in international business dealings.

Our journey gains momentum in Chapter 23, illuminating the relationship between supply chain visibility and trust in international markets. We scrutinize how blockchain's transparency becomes the bedrock for transforming supply chains, amplifying efficiency, and infusing trust into each link of the global business network. Chapters 24 and 25 unfold the complexities and solutions within cross-border contracts and regulatory compliance. The narrative guides us through the intricacies of smart contracts executing agreements seamlessly and the imperativeness of adhering to global regulatory frameworks. This exploration encapsulates the challenges of international business and the pivotal role of blockchain in surmounting these hurdles.

Chapters 26 and 27 open doors to innovative dimensions of global business. We traverse the realm where the tokenization of global assets becomes a beacon for efficiency in international transactions. Simultaneously, the spotlight is cast on the critical importance of identity management, ensuring the security and legitimacy of participants in the international business arena.

As we progress into Chapters 28 and 29, the narrative demonstrates the progressive features of multi-lingual smart contracts. Overcoming language barriers, these contracts become conduits for global collaborations. The emphasis then shifts to the strategic importance of international partnerships, laying the groundwork for synergies that transcend geographical boundaries.

The penultimate chapters, 30 and 31, delve into the strategic imperatives of talent acquisition and cross-border marketing. In a world where businesses extend their reach globally, acquiring the right talent and deploying effective marketing strategies become integral components of success.

Collectively, these chapters weave a rich ecosystem of insights, showcasing the profound impact of blockchain on the international business landscape. As we journey through the dance between technology and global commerce, the narrative encapsulates recurring themes of decentralization, security, transparency, and automation – qualities that align seamlessly with the demands of a globalized economy.

This introduction sets the stage for an immersive exploration of blockchain's role in navigating global frontiers. As we venture into the chapters ahead, the promise of uncovering new dimensions of blockchain's influence on the global stage beckons. The uncharted horizons await, promising deeper insights into the symbiotic relationship between blockchain technology and the ecosystem of international business.

Chapter 21: Global Payments and Remittances- Efficient Cross Border Transactions

Introduction

In the interconnected world of international business, the efficiency and reliability of global payment systems play a pivotal role. Here we look into the significance of these systems, exploring how blockchain technology revolutionizes cross-border transactions. As we navigate through the complexities of international finance, the cryptoverse emerges as a catalyst for seamless and efficient global payments.

Global payment systems are the lifeblood of international trade and commerce. Businesses, regardless of size, engage in transactions that traverse borders, necessitating a robust and swift payment infrastructure. The conventional systems, while effective, often grapple with challenges such as delays, high fees, and limited transparency.

Blockchain's Role in Revolutionizing Cross-Border Transactions

Enter blockchain technology — a disruptive force that redefines the landscape of global payments. The decentralized and distributed nature of blockchain eliminates the need for intermediaries, ushering in a new era of transparency, speed, and cost-efficiency.[89] Blockchain's impact on cross-border transactions is transformative, offering solutions to longstanding challenges. The technology provides a decentralized ledger where transactions are recorded in a secure and transparent manner. This not only minimizes the risk of fraud but also

---

[89] Staff (2023). "Cross-Border Payments and Blockchain Technology." LCX at https://www.lcx.com/cross-border-payments-and-blockchain-technology/.

ensures that transaction history is immutable and easily accessible.[90]

Smart contracts, a hallmark of blockchain, add another layer of efficiency to global payments. These self-executing contracts automatically enforce predefined terms when specific conditions are met.[91] In the context of international transactions, smart contracts streamline processes, reducing the need for manual intervention and mitigating the risk of errors.

Reducing Transaction Costs, Enhancing Speed, and Transparency

One of the persistent challenges in traditional global payment systems is the burden of high transaction costs. Blockchain, with its decentralized architecture, eliminates the need for multiple intermediaries and associated fees.[92] This cost-effective approach not only benefits businesses by reducing expenses but also democratizes access to international transactions.

Blockchain's distributed ledger ensures real-time updates on the status of transactions. This level of transparency is particularly crucial for businesses engaged in cross-border trade, providing them with a clear view of the entire payment process.[93] The removal of intermediaries also accelerates transaction settlement, enabling faster and more efficient cross-border payments.

---

[90] Staff (2023). "Blockchain for Cross-Border Payments." Science Soft at https://www.scnsoft.com/blockchain/cross-border-payments.

[91] *Id.*

[92] Harmse, Chris (2023). "Blockchain in Cross-Border payments: 2023 guide." Bvnk at https://www.bvnk.com/blog/blockchain-cross-border-payments.

[93] *Id.*

Challenges in Global Payments

While blockchain offers a promising solution, challenges persist. The volatility of cryptocurrencies, scalability issues, and regulatory uncertainties pose hurdles to widespread adoption.[94] Overcoming these challenges requires collaborative efforts from the industry, regulators, and technology developers.

Although a detailed discussion of SWIFT and Ripple/XRP is reserved for later chapters, it's essential to acknowledge their relevance. SWIFT, the incumbent in global payments, is facing competition from emerging blockchain-based alternatives.[95] Ripple's XRP, leveraging blockchain for cross-border payments, introduces new possibilities with its focus on speed and efficiency.

Conclusion: Navigating the Future of Cross-Border Transactions

Chapter 21 unravels the web of global payment systems, highlighting the transformative influence of blockchain technology. As we move forward in this exploration of the cryptoverse, the chapters ahead will delve deeper into specific programs, regulations, and advancements that shape the future of cross-border transactions. In the evolving landscape of international business, blockchain stands as a beacon of efficiency, promising a future where global payments are seamless, transparent, and accessible to all.

---

[94] Staff (2023). "How does Blockchain make Cross-Border Payments Better?" Crypto Council for Innovation at https://cryptoforinnovation.org/what-are-cross-border-payments-and-how-do-they-work/.

[95] *Id.*

Chapter 22: Trade Finance: Facilitation of International Trade

Introduction

> *Importance of International Trade: as the world becomes more interconnected 90% of business have some relation to international trade.*

Chapter 22 explores the expansive role of cryptocurrency in facilitating international trade, extending beyond its efficiency in cross-border transactions.[96] As we navigate the evolving landscape of global commerce, cryptocurrency emerges as a transformative force, redefining the s of trade, finance, and business relationships.

Cryptocurrency's Impact on International Trade

Cryptocurrency, with its decentralized and borderless nature, transcends the limitations of traditional financial systems.[97] Its impact on international trade is multi-faceted, influencing various aspects of commerce, from payments to supply chain management.

- Borderless Transactions and Reduced Friction
  - The inherent nature of cryptocurrency enables borderless transactions, eliminating the need for intermediaries and reducing friction in international trade. Traditional banking systems often introduce delays, high fees, and complex

---

[96] Staff (2023). "Blockchain for Global Trade and Commerce." Consensys at https://consensys.io/blockchain-use-cases/global-trade-and-commerce.

[97] UNCTAD (2024). "Blockchain for facilitating trade and enhancing competitiveness." At https://unctad.org/project/blockchains-facilitating-trade-and-enhancing-competitiveness.

processes.[98] Cryptocurrency, as a medium of exchange, facilitates direct, peer-to-peer transactions, streamlining the payment process and reducing associated costs.

- Enhanced Security and Trust
    - The cryptographic principles underlying cryptocurrency transactions provide a heightened level of security. The decentralized and tamper-resistant nature of blockchain technology ensures the integrity and immutability of transaction records. This security aspect fosters trust among international trading partners, especially in regions where concerns about fraud and lack of transparency may hinder business relationships.[99]
- Decentralized Finance (DeFi) in International Trade
    - The emergence of decentralized finance (DeFi) platforms powered by cryptocurrency introduces innovative financial instruments for international trade. Smart contracts, decentralized exchanges, and blockchain-based lending systems offer new avenues for financing trade transactions.[100] This decentralized approach enables businesses to access financial services without traditional banking

---

[98] Ganne, Emmanuelle (2023). "Can Blockchain Revolutionize International Trade." World Trade Organization at https://www.wto.org/english/res_e/publications_e/blockchainrev18_e.htm.

[99] Barafani, Magdalena; Garcia, Pablo & Rozemberg, Ricardo (2021). "Blockchain Technology: a New Opportunity for International Trade." IDB at https://blogs.iadb.org/integration-trade/en/blockchain-technology-a-new-opportunity-for-international-trade/.

[100] Consensys, *supra*.

intermediaries, promoting financial inclusion in the global market.

- Transparent and Traceable Supply Chains
    - Blockchain, the underlying technology of many cryptocurrencies, plays a pivotal role in enhancing the transparency and traceability of supply chains. The ability to record every step of a product's journey on an immutable ledger ensures authenticity and compliance with international standards.[101] This transparency is particularly crucial in industries where provenance and adherence to regulations are paramount, such as food and pharmaceuticals.
- Mitigating Currency Risks
    - International trade often involves dealing with multiple currencies, exposing businesses to currency exchange risks.[102] Cryptocurrencies, being inherently global and not tied to any specific country, provide a potential solution to mitigate these risks. Stablecoins, pegged to traditional fiat currencies, offer a bridge between the benefits of cryptocurrency and the stability of established monetary systems.[103]

---

[101] UNCTAD, *supra*.

[102] UNCTAD (2024A). "Global Report on Blockchain and its implications on Trade Facilitation Performance." UN at https://unctad.org/publication/global-report-blockchain-and-its-implications-trade-facilitation-performance.

[103] UNECE White Paper at https://unece.org/fileadmin/DAM/cefact/GuidanceMaterials/WhitePaperBlockchain.pdf.

Conclusion: Navigating the Future of International Trade

Chapter 22 underscores the transformative influence of cryptocurrency on international trade, extending beyond its role in cross-border transactions. From reducing friction in payments to fostering transparent supply chains, cryptocurrency reshapes the s of global commerce. As businesses and nations adapt to the cryptoverse, the potential for innovation in trade practices, financing, and supply chain management becomes increasingly evident. In the chapters ahead, we will delve deeper into specific use cases, challenges, and advancements, providing a comprehensive understanding of how cryptocurrency navigates the landscape of international trade.

Chapter 23: Supply chain Visibility- Building Trust Across Borders

Introduction

Chapter 23 delves into the pivotal role of supply chain visibility in fostering trust within international markets.[104] In an era where global commerce demands transparency and accountability, the ability to trace and verify the journey of products becomes a foundation for building trust among trading partners. This chapter explores the multifaceted impact of supply chain visibility, examining its implications for businesses, consumers, and the overall integrity of international trade.[105]

Modern supply chains are networks that span the globe, involving various stakeholders, from manufacturers to distributors and retailers. The complexity of these chains often leads to challenges such as inefficiencies, delays, and a lack of transparency. Supply chain visibility addresses these challenges by providing real-time insights into the movement and status of goods throughout the entire supply network.[106] Supply chain

---

[104] Staff (2023). "The Use of Blockchain in Supply Chain for Security and Visibility." TIBC at https://www.tibco.com/reference-center/the-use-of-blockchain-in-supply-chain-for-security-and-visibility.

[105] Rogerson, Michael & Parry, Glenn (2020). "Blockchain: Case Studies in Food Supply chain visibility." Supply chain Management Journal at https://www.emerald.com/insight/content/doi/10.1108/SCM-08-2019-0300/full/html?casa_token=m2exJG2QYfEAAAAA:TDuoIl68BqrFvJWqA0KLCkK-XFExJKuHRhoTWQkOOqbJ_W1BEWs807sVXBYpRK2tRJeuabfYknSypiWSSakPk64keZumd6WdL8MBUUGBEKQN5eCdLFfh.

[106] Hamledari, Hesam & Fischer, Martin (2021). "Measuring the impact of Blockchain and smart contracts on construction supply chain visibility." Advanced Engineering Informatics (v 50) at https://www.sciencedirect.com/science/article/pii/S147403462100196 8?casa_token=E7c2drhScYcAAAAA:tDGNJg9XAmL3r-

visibility leverages technologies such as blockchain, IoT (Internet of Things), and advanced analytics to enable real-time tracking of goods. This transparency ensures that each step of the supply chain is documented and accessible to relevant parties. As a result, businesses can gain immediate insights into inventory levels, production milestones, and delivery schedules.

Enhancing Trust in International Trade

In international trade, where transactions span continents and involve diverse regulatory frameworks, trust is a critical currency. Supply chain visibility acts as the foundation of this trust by providing stakeholders with a clear, unambiguous view of the entire process.[107] Businesses can showcase their commitment to transparency, thereby instilling confidence in partners and customers alike.

Visibility into the supply chain allows businesses to enforce quality standards and ensure compliance with industry regulations.[108] Whether dealing with perishable goods or regulated products, the ability to trace each item's origin, storage conditions, and handling practices enhances the credibility of businesses in international markets.[109] This, in turn, contributes to a culture of trust among trading partners.[110]

Beyond transparency, supply chain visibility contributes to the efficiency of international trade. Predictive analytics powered by real-time data can anticipate potential disruptions, enabling

---

nQ0LoBFn0Yv8BVyGE2IJnail0xEfYegYDJqkejYvl-0ADmbgNySz8-UEdc7KA.

107 Sunmola, Funlade & Apeji, Uje (2020). "Blockchain Characteristics for Supply Chain Visibility." NA International Conference on Industrial Engineering and Operations Management at http://www.ieomsociety.org/detroit2020/papers/660.pdf.

108 Rogerson & Parry, *supra*.

109 *Id*.

110 TIBC, *supra*.

proactive risk mitigation strategies.[111] Timely responses to challenges such as natural disasters, geopolitical events, or unexpected demand fluctuations further strengthen the reliability of businesses engaged in cross-border trade.

Challenges and Opportunities

While the benefits of supply chain visibility are substantial, implementing such systems comes with its own set of challenges. Integrating diverse technologies, ensuring data accuracy, and fostering collaboration among supply chain participants require careful planning and investment.[112] However, overcoming these challenges opens the door to a more resilient and trustworthy international trade ecosystem.

Supply chain visibility not only addresses existing issues but also presents opportunities for innovation and differentiation.[113] Businesses that embrace advanced technologies to enhance transparency demonstrate a commitment to excellence and customer satisfaction. In the competitive landscape of international trade, this commitment becomes a distinguishing factor.

Conclusion

In summary, By highlighting the transformative impact of supply chain visibility on international trade we open the discussion to new opportunities. As businesses navigate the complexities of global commerce, the ability to provide visibility into the supply chain becomes a strategic imperative. From bolstering trust to

---

[111] Li, Zhijie; Wu, Haoyan; King, Brian; Miled, Zina; Wassick, John & Tazelaar, Jeffery (2018). "A Hybrid Blockchain Ledger for Supply Chain Viability." IEEE at https://ieeexplore.ieee.org/abstract/document/8452028?casa_token=0QSwxslu4HgAAAAA:Z0Ee6_tvXMvG0QFyceh65kmfpxelgsBZN0m8rEVEOPVVUWn92JY0rXs09Zfv7Dt26ESZGQSvBw.

[112] *Id.*

[113] Sunmola, Funlade & Apeji, Uje, *supra*.

fostering efficiency and innovation, supply chain visibility emerges as a catalyst for shaping the future of international trade. The subsequent chapters will further explore specific use cases, emerging technologies, and the evolving landscape of supply chain s within the cryptoverse.

Chapter 24: Cross Borders Contracts- Smart Legal Agreements

Introduction:

In the complex landscape of international business, cross-border contracts are pivotal components, defining agreements between parties spanning different countries. However, the intricacies of these contracts often give rise to challenges, including issues of trust, legal enforcement, and efficient execution. In this chapter, we explore the hurdles associated with cross-border contracts and delve into how blockchain technology and cryptocurrency offer innovative solutions, promising increased transparency, security, and reliability in international business dealings.

Challenges in Cross-Border Contracts:

1. *Legal Diversities:* The legal frameworks governing contracts vary significantly across borders, leading to complexities in understanding and enforcing agreements.[114] These differences can result in disputes, delays, and increased costs for businesses.[115]
2. *Trust Deficits:* Establishing trust between parties located in different jurisdictions can be challenging.[116] Lack of transparency and concerns about the reliability of contractual counterparts often hinder smooth collaboration.

---

[114] Lehmann, Matthias (2021). "National Blockchain Laws as a Threat to Capital Markets integration." Revue de droit uniforme at https://academic.oup.com/ulr/article/26/1/148/6314582.

[115] Hacker, Philipp; Lianos, Loannis; Dimitropolous, Georgios & Eich, Stefan (2019). Regulating Blockchain: Techno-Social and Legal Challenges. Oxford University Press: Oxford.

[116] Kundu, Debasish (2022). "Blockchain and Trust in a Smart City." National Institute of Urban Affairs at https://journals.sagepub.com/doi/abs/10.1177/0975425319832392.

3. *Payment Inefficiencies:* Traditional cross-border transactions are plagued by inefficiencies, involving multiple intermediaries, delays, and high transaction costs.[117] These issues impact the timely execution of contractual obligations.
4. *Contractual Disputes:* Resolving disputes in cross-border contracts can be a protracted and expensive process.[118] The involvement of legal systems from different jurisdictions adds layers of complexity, slowing down the resolution process.

Blockchain Solutions:

1. *Smart Contracts:* Utilizing smart contracts, self-executing agreements with code-enforced terms, brings automation and transparency to cross-border contracts.[119] Smart contracts operate on blockchain networks, ensuring tamper-resistant execution when predefined conditions are met.
2. *Immutable Record-Keeping:* Blockchain's decentralized and immutable nature ensures that once a contract is recorded on the blockchain, its content remains unchanged.[120] This feature adds a layer of security and eliminates concerns about fraudulent alterations.
3. *Transparency and Traceability:* Blockchain provides a transparent and traceable ledger of all transactions

---

[117] Staff (2023) BIG at https://almondfintech.com/wp-content/uploads/2023/03/Cross-Border-Payments-Challenges-Impact-and-Solutions.pdf.

[118] Giambrone Law (2024). "Types of Cross Border disputes and options for solving them." At https://www.giambronelaw.com/site/advice/dispute-resolution/commercial-cross-border-disputes/types-of-commercial-cross-border-disputes/.

[119] Lehmann, *supra.*

[120] Hacker et al, *supra*

related to a contract.[121] This visibility enhances trust between parties, allowing them to verify the history and progression of the agreement.

4. *Decentralized Identity Verification:* Blockchain facilitates decentralized identity verification, streamlining the onboarding process and addressing trust deficits. Parties can rely on verified; immutable identities recorded on the blockchain.[122]

Cryptocurrency Solutions:

1. *Efficient Cross-Border Payments:* Cryptocurrencies, like Bitcoin and stablecoins, offer faster and cost-effective alternatives to traditional banking systems for cross-border payments. By bypassing intermediaries, businesses can execute payments promptly.
2. *Currency Agnosticism:* Cryptocurrencies are not bound by specific national currencies, reducing the impact of currency fluctuations. This feature allows parties to denominate contracts in cryptocurrencies, mitigating risks associated with volatile fiat currencies.
3. *Borderless Transactions:* Cryptocurrencies operate on a global scale, providing a borderless financial infrastructure.[123] This allows parties in different countries to transact seamlessly without the constraints of traditional financial systems.

Conclusion:

In conclusion, the challenges inherent in cross-border contracts necessitate innovative solutions to facilitate smoother international business transactions. Blockchain technology, with its emphasis on transparency, security, and automation through

---

[121] Kundu, *supra*.
[122] Giambrone Law, *supra*.
[123] Smithmyer et al, *supra*.

smart contracts, presents a promising avenue for addressing these challenges. Coupled with the efficiency and borderless nature of cryptocurrencies, businesses can navigate the complexities of cross-border contracts with increased confidence and reliability, ushering in a new era of international collaboration.

Chapter 25: Regulatory Compliance-Navigating International Regulations

Introduction:

Regulatory compliance stands as a cornerstone in the ever-evolving landscape of blockchain and cryptocurrency. As these technologies disrupt traditional financial systems and redefine business paradigms, adherence to regulatory frameworks becomes imperative. Here we analyze the pivotal role of regulatory compliance, exploring its importance in fostering legitimacy, mitigating risks, and ensuring the sustainable growth of blockchain and cryptocurrency ecosystems.

Navigating the Regulatory Landscape:

1. *Legitimacy and Trust:* Regulatory compliance establishes a foundation of legitimacy for blockchain and cryptocurrency projects.[124] Adhering to established rules and regulations fosters trust among users, investors, and stakeholders, crucial for the sustained growth of the industry.
2. *Mitigating Legal Risks:* The nascent nature of the blockchain industry often intersects with existing legal frameworks. Regulatory compliance acts as a shield, protecting projects from legal risks and potential litigations. Clear compliance frameworks reduce uncertainties and instill confidence in both businesses and users.
3. *Investor Protection:* Regulatory compliance safeguards the interests of investors by ensuring that projects operate within defined legal boundaries.[125] This

---

[124] Staff (2017). "7 Regulatory Challenges facing Blockchain." BBVA at https://www.bbva.com/en/7-regulatory-challenges-facing-blockchain/.
[125] *Id.*

protection is vital for attracting institutional investors who, in turn, contribute to the industry's maturation.

4. *Preventing Financial Crimes:* Adherence to anti-money laundering (AML) and know your customer (KYC) regulations is integral in preventing illicit activities within blockchain and cryptocurrency networks. Compliance measures serve as deterrents, contributing to a secure and transparent financial environment.

Challenges in Regulatory Compliance:

1. *Global Variances:* The absence of uniform global regulations poses challenges for projects operating in multiple jurisdictions.[126] Navigating through diverse regulatory landscapes requires flexibility and a deep understanding of regional compliance requirements.
2. *Evolving Standards:* Regulatory standards for blockchain and cryptocurrency are continuously evolving.[127] Staying abreast of changing regulations is a perpetual challenge, demanding proactive engagement and a commitment to adapt to emerging compliance norms.
3. *Balancing Innovation and Compliance:* Striking a balance between fostering innovation and adhering to regulatory guidelines is an ongoing challenge.[128] Regulatory bodies aim to safeguard financial systems, while innovators seek to push the boundaries of what is possible within the realm of blockchain and cryptocurrency.

Importance of Self-Regulation:

1. *Industry Accountability:* Self-regulatory initiatives within the blockchain and cryptocurrency space demonstrate

---

[126] Smithmyer *et al, supra.*

[127] *Id.*

[128] *Id.*

industry commitment to ethical practices. These initiatives, such as codes of conduct and best practices, showcase a collective effort to maintain high standards of integrity.

2. *Enhancing Credibility:* Proactive self-regulation enhances the credibility of the industry.[129] Projects that voluntarily adopt and adhere to self-regulatory measures signal their commitment to responsible conduct, fostering a positive reputation within the broader business ecosystem.

Conclusion:

In conclusion, regulatory compliance is not merely a legal obligation but a fundamental enabler of the sustained growth and acceptance of blockchain and cryptocurrency. The importance of navigating complex regulatory landscapes cannot be overstated. As the industry matures, a collaborative approach involving regulatory bodies, industry participants, and self-regulatory initiatives will be crucial in establishing a robust framework that balances innovation with legal adherence. By embracing and championing regulatory compliance, the blockchain and cryptocurrency industry can continue its journey toward broader acceptance, mainstream integration, and long-term viability.

[129] Smithmyer, Christopher (2023A). "Digital Exchequer." Elite Exclusivity: Detroit.

Chapter 26: Tokenization of Global Assets-Accessing Global Investment

Introduction:

The tokenization of global assets represents a transformative shift in the landscape of international business. This chapter delves into the profound implications of tokenizing assets on a global scale, exploring how this innovative approach enhances liquidity, efficiency, and accessibility in the realm of international trade. Tokenization involves representing real-world assets, such as real estate, stocks, or commodities, as digital tokens on a blockchain. These tokens, often divisible and transferable, enable fractional ownership, creating a new paradigm for the global asset landscape.

Enhancing Liquidity:

1. *Fractional Ownership:* Tokenization allows assets to be divided into smaller, tradable units. This fractional ownership democratizes access to high-value assets, unlocking liquidity and broadening investment opportunities.[130]
2. *24/7 Trading:* The blockchain operates seamlessly around the clock, enabling continuous trading of tokenized assets without the limitations imposed by traditional financial markets' operating hours. This aspect enhances market efficiency and accessibility for global participants.[131]

---

[130] CFI (2024). "Liquidity in Cryptocurrency." At https://corporatefinanceinstitute.com/resources/cryptocurrency/liquidity-in-cryptocurrency/.

[131] *Id.*

Efficiency in Transactions:

1. *Smart Contracts and Automation:* Tokenized assets often leverage smart contracts for automated execution of predefined conditions. This reduces the need for intermediaries, streamlining transaction processes and minimizing delays associated with traditional settlement systems.
2. *Cross-Border Transactions:* Tokenization facilitates cross-border transactions by eliminating barriers like currency conversions and protracted settlement times. This efficiency is particularly valuable in international business, where time-sensitive transactions are common.

Accessibility and Inclusivity:

1. *Global Investment Opportunities:* Tokenization transcends geographical barriers, providing investors worldwide with access to a diverse array of global assets. This inclusivity fosters a more interconnected and international investment landscape.[132]
2. *Reduced Entry Barriers:* Fractional ownership and lower transaction costs associated with tokenized assets reduce entry barriers for retail investors. This inclusiveness democratizes the investment landscape, allowing a more diverse pool of participants.[133]

Challenges and Considerations:

1. *Regulatory Frameworks:* The tokenization of global assets introduces new regulatory considerations.

---

[132] Gulen, Karem (2023). "Exploring the most efficient blockchain technology." Data Economy at https://dataconomy.com/2023/05/24/most-efficient-blockchain-technology/.

[133] *Id.*

Harmonizing regulatory frameworks globally is crucial to ensure compliance and foster widespread adoption.[134]

2. *Security Concerns:* While blockchain offers enhanced security, tokenization platforms must address potential vulnerabilities and ensure robust security measures to protect investors and asset issuers.

Conclusion:

In conclusion, the tokenization of global assets emerges as a powerful force reshaping international business s. By enhancing liquidity, streamlining transactions, and fostering accessibility, tokenization creates a more inclusive and efficient global asset ecosystem. While challenges exist, the potential benefits for international trade, investment, and financial inclusivity position tokenization as a pivotal innovation with far-reaching implications. As businesses and regulators navigate this transformative landscape, the tokenization of global assets is poised to play a central role in shaping the future of international commerce.

---

[134] Giambrone Law, *supra*.

Chapter 27: Multi-Lingual Smart Contracts-Bridging language barriers

Introduction

Language diversity has been a longstanding challenge in international business and finance. This chapter explores the innovative integration of multilingual smart contracts, examining how these solutions effectively address language barriers within the fintech landscape.

Multilingual smart contracts represent a groundbreaking advancement in smart contract technology. Unlike traditional contracts written in a specific language, these contracts can be executed in multiple languages, overcoming linguistic barriers that have hindered global financial interactions.

Breaking Down Language Barriers:

1. *Smart Contracts as Code:* Multilingual smart contracts leverage code as their language, making them inherently adaptable to various linguistic interfaces.[135] This ensures that participants can engage with contracts using their preferred language without compromising functionality.
2. *Translation Mechanisms:* These contracts incorporate advanced translation mechanisms, allowing the code to be translated seamlessly into multiple languages.[136] This ensures that the contractual terms are accurately understood by all parties involved, regardless of their linguistic background.

---

[135] Staff (2023). "How can cryptocurrency content translation help break down language barriers." Optimational at https://www.optimational.com/blog/cryptocurrency-content-translation-overcome-language-barriers/.

[136] *Id.*

Facilitating Cross-Border Transactions:

1. *Global Accessibility:* Multilingual smart contracts enhance the accessibility of financial services and transactions on a global scale. Participants from diverse linguistic backgrounds can engage in smart contract-based agreements without language becoming a barrier.
2. *Legal Clarity:* Overcoming language barriers in contractual agreements is crucial for legal clarity. Multilingual smart contracts contribute to reducing ambiguity, ensuring that all parties have a clear and consistent understanding of the contractual terms.

Operational Efficiency:

1. *Automation and Efficiency:* The automated nature of smart contracts combined with multilingual capabilities streamlines operational processes.[137] Parties can execute agreements without the need for manual translation, reducing delays and enhancing overall efficiency.
2. *Real-Time Communication:* These contracts enable real-time communication and updates in multiple languages, fostering transparent and efficient interactions. This is particularly beneficial in financial environments where quick responses are essential.

1. *Ensuring Accuracy:* The accurate translation of complex financial terms and conditions poses a challenge. Implementing robust translation algorithms is crucial to maintain precision in multilingual smart contracts.
2. *Regulatory Compliance:* Navigating diverse regulatory landscapes becomes more complex when dealing with

---

[137] McClingtock, Matt (2020). "Blockchain Workflow Automation: Why you should embrace it." Processmaker at https://www.processmaker.com/blog/blockchain-workflow-automation-why-you-should-embrace-it/.

multiple languages. Ensuring compliance with regional regulations is a key consideration in the development and deployment of multilingual smart contracts.

Conclusion:

In conclusion, the integration of multilingual smart contracts marks a pivotal advancement in fintech, addressing the longstanding challenge of language barriers in global financial interactions. By breaking down linguistic obstacles, these contracts enhance accessibility, legal clarity, and operational efficiency. While challenges exist, the potential benefits for fostering international collaboration and inclusivity position multilingual smart contracts as a transformative solution in the evolving landscape of fintech. As businesses embrace this innovation, the ability to conduct seamless and multilingual transactions opens new frontiers in the global financial ecosystem.

Chapter 28: International Partnerships- Collaborating Across Borders

Introduction:

In the interconnected landscape of fintech, international partnerships have emerged as a strategic imperative for businesses seeking global relevance and influence. This chapter delves into the profound importance of forging and nurturing international collaborations within the fintech sector.

The s of International Partnerships:

1. *Expanding Market Reach:* International partnerships open doors to new markets, enabling fintech companies to broaden their geographic footprint. Collaborations facilitate market entry by leveraging the knowledge and networks of local partners.
2. *Diverse Expertise and Innovation:* Partnering with organizations from different regions brings a wealth of diverse expertise and perspectives. This diversity fosters innovation, as cross-cultural collaborations often lead to creative solutions and novel approaches to industry challenges.
3. *Risk Mitigation:* International partnerships can act as a risk mitigation strategy. By diversifying operations across multiple regions, fintech companies can reduce their exposure to geopolitical, economic, and regulatory risks associated with a single market.

Strategic Alliances in Fintech:

1. *Financial Inclusion Initiatives:* Collaborations between fintech firms and international organizations often focus on advancing financial inclusion.[138] By combining

[138] Klus, Milan & Lohwasser, Todor (2019). "Strategic Alliances between banks and fintechs for digital innovation: Motives to collaborate and

resources and expertise, these alliances work towards providing accessible and inclusive financial services globally.

2. *Technology Exchange Programs:* International partnerships facilitate the exchange of technological advancements. Fintech firms can leverage each other's innovations, accelerating the development and adoption of cutting-edge technologies across borders.[139]
3. *Regulatory Alignment:* Establishing partnerships with organizations in regions with aligned regulatory frameworks streamlines compliance efforts.[140] This is particularly crucial in fintech, where regulatory nuances vary significantly between jurisdictions.

Navigating Cultural Nuances:

1. *Understanding Local Preferences:* International partnerships require a nuanced understanding of local consumer preferences and behaviors. Fintech companies need to adapt their offerings to cater to diverse cultural expectations, ensuring resonance with the target audience.[141]
2. *Communication Strategies:* Effective communication becomes paramount in cross-cultural collaborations. Fintech firms must employ strategies that bridge linguistic and cultural gaps, fostering clear and concise communication among partners.

---

types of interaction." Journal of Entrepreneurial Finance at https://digitalcommons.pepperdine.edu/cgi/viewcontent.cgi?article=1346&context=jef.

[139] Staff (2023). "5 Bank and Fintech Partnership Models and their benefits." Mineral Tree at https://www.mineraltree.com/blog/3-types-fintech-partnerships/.

[140] Klus & Lohwaser, *supra*.

[141] Smithmyer (2023A), *supra*.

1. *Regulatory Hurdles:* Divergent regulatory environments can pose challenges in international collaborations. Fintech companies must navigate complex regulatory landscapes to ensure compliance and seamless operations.
2. *Cultural Misalignment:* Misunderstandings stemming from cultural differences can impede the success of international partnerships. Developing cultural intelligence and fostering an inclusive environment are critical to overcoming these challenges.

The Future Landscape of Fintech Collaborations:

1. *Blockchain Consortia:* Fintech companies are increasingly forming blockchain consortia, uniting forces to explore and implement distributed ledger technologies. These consortia often involve members from different parts of the world collaborating on blockchain solutions.
2. *Global Innovation Hubs:* The rise of global innovation hubs brings together fintech startups, industry leaders, and regulatory bodies from various regions. These hubs serve as focal points for international collaboration, driving collective efforts towards industry advancements.

Conclusion:

International partnerships stand as bulwarks in the evolving narrative of fintech. As the industry continues to transcend borders, collaborative efforts become instrumental in driving innovation, expanding market presence, and navigating the complexities of a globalized financial landscape. The ability to forge meaningful partnerships across international boundaries not only positions fintech companies for sustained success but also contributes to the collective advancement of the global fintech ecosystem. Embracing the challenges and opportunities inherent in international collaborations is essential for those seeking to thrive in the and interconnected world of fintech.

Chapter 29: Talent Acquisition- Building International Teams

Introduction:

In the landscape of blockchain technology, talent acquisition stands as a critical cornerstone for businesses aiming to build and sustain impactful blockchain programs. This chapter explores the paramount importance of acquiring skilled professionals and delves into strategic methodologies for recruiting talent in the realm of blockchain development.

The Crucial Role of Talent in Blockchain Programs:

1. **Technical Expertise:** Building successful blockchain programs demands a deep reservoir of technical expertise. Skilled professionals proficient in distributed ledger technologies, smart contract development, cryptography, and consensus algorithms are essential assets.
2. **Innovation Catalysts:** Blockchain is at the forefront of technological innovation. Talented individuals bring creativity, ingenuity, and a forward-thinking mindset, serving as catalysts for developing groundbreaking solutions and pushing the boundaries of what is possible.
3. **Problem Solvers:** Blockchain programs often grapple with complex challenges. A talented team adept at creative problem-solving is instrumental in overcoming obstacles, ensuring the smooth development and implementation of blockchain solutions.

Strategic Methodologies for Talent Acquisition:[142]

1. **Internal Skill Development:** Businesses can foster talent acquisition by investing in internal skill development programs. Training existing employees on blockchain technologies helps in cultivating an in-house pool of blockchain experts.
2. **Collaboration with Educational Institutions:** Establishing partnerships with educational institutions, universities, and research centers provides a pipeline for fresh talent. Companies can engage in mentorship programs, sponsorships, and internships to identify and nurture emerging talent.
3. **Active Participation in Industry Events:** Networking plays a pivotal role in talent acquisition. Actively participating in industry conferences, meetups, and events allows businesses to connect with professionals passionate about blockchain, creating avenues for recruitment.
4. **Global Talent Search:** The distributed nature of blockchain technology allows companies to cast a wide net for talent globally. Remote work opportunities and global talent searches enable businesses to access a diverse pool of skilled professionals. This requires competitive payment packages.

Navigating Challenges in Talent Acquisition:

1. **High Demand and Short Supply:** The demand for blockchain talent often outstrips the available supply. Businesses must be proactive in their recruitment

[142] Weinstein, Gary (2024). "Crypto talent war for blockchain supremacy: The US versus the world." Forbes at https://www.forbes.com/sites/digital-assets/2023/05/30/crypto-talent-war-for-blockchain-supremacy-the-us-versus-the-world/?sh=2ba14dd97948.

strategies, building relationships with professionals before the need becomes critical.

2. **Evolution of Skill Requirements:** The blockchain space evolves rapidly, leading to continuous shifts in skill requirements. Companies should adopt flexible recruitment strategies and prioritize candidates with a capacity for continuous learning and adaptability.

Nurturing a Positive Organizational Culture:

1. **Fostering Collaboration:** A collaborative and inclusive organizational culture attracts and retains top talent. Blockchain development often involves interdisciplinary teams, emphasizing the importance of a culture that fosters collaboration.
2. **Professional Development Opportunities:** Providing avenues for professional growth and development enhances employee satisfaction and retention. Blockchain professionals value organizations that invest in their continuous learning and skill enhancement.

Conclusion:

In the journey of building blockchain programs, talent acquisition emerges as a strategic imperative. The success of blockchain initiatives hinges on the caliber of individuals steering the development and implementation processes. By adopting proactive recruitment strategies, fostering a positive organizational culture, and addressing the unique challenges of the blockchain talent landscape, businesses can position themselves as leaders in the ever-evolving realm of blockchain technology. The pursuit of top-tier talent is not just an investment in individuals but a strategic commitment to the success and longevity of blockchain programs in the competitive global landscape.

Chapter 30: Cross Border Marketing-Reaching Global Audiences

Introduction:

In the era of globalization, businesses are increasingly expanding their reach beyond borders, necessitating a nuanced approach to marketing that transcends geographical constraints. This chapter delves into the critical significance of cross-border marketing and explores how the blockchain, with its inherent features, provides a transformative advantage in navigating the complexities of international marketing.

The Imperative of Cross-Border Marketing:[143]

1. **Expanding Customer Base:** Cross-border marketing enables businesses to tap into diverse markets, unlocking new opportunities and expanding their customer base. It allows companies to reach untapped demographics and demographics with different cultural nuances.
2. **Mitigating Market Saturation:** In highly competitive industries, penetrating new markets becomes essential to combat market saturation. Cross-border marketing provides a strategic avenue for growth, ensuring companies remain agile and adaptable to changing markets.
3. **Diversification of Revenue Streams:** Relying solely on domestic markets poses inherent risks. Cross-border marketing allows businesses to diversify their revenue streams, reducing dependence on a single market and enhancing overall business resilience.

---

[143] Sandor, Krisztian (2023). "Fresh Money flows to crypto as stablecoin market expands after 18-month downtrend." Coindesk at https://www.coindesk.com/markets/2023/12/05/fresh-money-flows-to-crypto-as-stablecoin-market-expands-after-15-years-downtrend/.

Challenges in Cross-Border Marketing:[144]

1. **Cultural and Linguistic Barriers:** Understanding and navigating diverse cultural and linguistic landscapes can be challenging. Tailoring marketing strategies to resonate with different cultural sensibilities is crucial for success in international markets.
2. **Regulatory Compliance:** Varied regulatory frameworks across countries demand meticulous attention to compliance. Navigating legal requirements and ensuring adherence to regulations in each market is a complex yet critical aspect of cross-border marketing.

The Blockchain Advantage in Cross-Border Marketing:

1. **Enhanced Security and Transparency**: Blockchain's decentralized, and tamper-resistant nature ensures enhanced security and transparency in marketing transactions. This instills confidence among international consumers and partners, mitigating concerns related to data security and fraudulent activities.
2. **Smart Contracts for Efficient Agreements**: Smart contracts automate and streamline agreements in marketing collaborations. They facilitate trust by executing predefined actions automatically when conditions are met, reducing the need for intermediaries, and ensuring efficient and secure collaborations across borders.
3. **Tokenization of Assets:** The tokenization of assets on the blockchain provides a novel approach to cross-border marketing. It allows businesses to create and distribute digital tokens that represent ownership or

---

[144] Pitney Bowes at https://www.pitneybowes.com/us/blog/8-common-challenges-of-cross-border-ecommerce-strategy-growth.html,

access rights, fostering a new paradigm for incentivizing and engaging international customers.

4. Decentralized Marketing Platforms: Blockchain enables the development of decentralized marketing platforms that connect businesses directly with consumers. This disintermediation ensures cost-effectiveness, transparency, and a direct line of communication, streamlining cross-border marketing efforts.

Case Study- VeChain in Supply Chain Marketing-

VeChain: Revolutionizing Supply Chain Marketing with Transparency and Trust

In today's hyper-connected world, consumers are increasingly demanding transparency and ethical practices from the brands they support. This is especially true in the realm of supply chains, where concerns about sustainability, authenticity, and fair labor practices are often shrouded in opacity. Enter VeChain, a blockchain-powered platform that is revolutionizing the way companies manage their supply chains and engage with their customers through transparent and trust-enhancing marketing strategies.

VeChain's Core Value Proposition:

VeChain leverages blockchain technology to create an immutable and transparent record of every step a product takes throughout its journey, from raw materials to the consumer's doorstep. This record, accessible via QR codes or embedded chips, empowers brands to tell the story of their products in a compelling and verifiable way. Consumers can scan these codes to access information such as provenance, production methods, sustainability practices, and even ethical sourcing initiatives.

Enhancing Marketing Strategies with VeChain:

VeChain's impact on supply chain marketing is multifaceted:

- **Building Brand Trust:** By providing verifiable proof of their claims, brands can build trust and loyalty with consumers who value transparency and ethical practices. This can lead to increased brand advocacy and positive word-of-mouth marketing.
- **Engaging Storytelling:** VeChain enables brands to go beyond traditional marketing narratives and tell the authentic story of their products. Consumers can connect with the human faces behind the products, the communities involved in their creation, and the positive impact they make on the environment and society.
- **Personalization and Engagement:** VeChain allows brands to personalize marketing campaigns based on individual consumer preferences and purchase history. By analyzing on-chain data, brands can target consumers with relevant information and offers, fostering deeper engagement and loyalty.
- **Fighting Counterfeiting:** VeChain's secure blockchain technology makes it virtually impossible to counterfeit products, protecting brands from reputational damage and lost revenue. This translates to a more trusted and valuable brand image for consumers.

Example Studies: VeChain in Action

- **BMW:** The luxury car manufacturer partnered with VeChain to track the production and distribution of its electric vehicles, showcasing the sustainable practices and ethical sourcing of materials used in their creation. This transparency campaign resonated with environmentally conscious consumers and boosted brand image.[145]

---

[145] Kiguru, Jiohn (2023). "VeChain to lead data management revolution due to partnerships with BYD, Renault & BMW." Crypto News Flash at https://www.crypto-news-flash.com/vechain-to-lead-data-

- **Walmart China:** To address concerns about food safety and provenance, Walmart China implemented VeChain to track the journey of its fresh produce. Consumers could scan QR codes on fruits and vegetables to access information about their origin, farming practices, and freshness, leading to increased trust and sales.[146]
- **DNV GL:** The sustainability certification body uses VeChain to track and verify the carbon footprint of products throughout their lifecycle. This data-driven approach empowers brands to communicate their environmental efforts effectively and attract eco-conscious consumers.

The Future of Supply Chain Marketing with VeChain:

As consumers continue to demand greater transparency and ethical practices, VeChain's role in supply chain marketing is poised to grow even further. We can expect to see new and innovative applications of the platform, such as:

- **Direct-to-consumer engagement:** Blockchain-based loyalty programs and rewards systems can incentivize consumers to engage with brands directly, fostering stronger relationships and brand loyalty.
- **Gamified experiences:** Interactive AR/VR experiences powered by VeChain data can allow consumers to virtually explore farms, factories, and communities involved in product creation, further deepening their connection with the brand.

---

management-revolution-due-partnerships-with-byd-renault-bmw-report/.

[146] Karunanidhi, Vignesh (2023). "VeChain and Walmart China team up for Blockchain Traceability." WatcherGuru at https://watcher.guru/news/vechain-and-walmart-china-team-up-for-blockchain-traceability.

- **Data-driven sustainability initiatives:** VeChain's ability to track and analyze environmental data can help brands develop and implement more effective sustainability strategies, attracting consumers who value eco-conscious practices.[147]

Conclusion of Case Study:
VeChain is not just a technological innovation; it represents a fundamental shift in how brands connect with consumers in the age of transparency and trust. By harnessing the power of blockchain, VeChain empowers brands to tell the authentic story of their products, build deeper relationships with their customers, and ultimately, create a more sustainable and ethical future for the global supply chain.

Conclusion:

In the landscape of cross-border marketing, the blockchain emerges as a transformative force, addressing challenges and unlocking new possibilities. Its decentralized, secure, and transparent nature aligns seamlessly with the requirements of international marketing, offering a strategic advantage to businesses aiming to transcend borders. As companies navigate the complexities of cultural diversity, regulatory landscapes, and evolving markets, integrating blockchain technology into cross-border marketing strategies becomes not just a choice but a strategic imperative for sustained success in the global marketplace.

[147] Staff (2023). "DNV GL and VeChain to showcase My Story ™ at Viva Technology." DNV at https://www.dnv.com/news/dnv-gl-and-vechain-to-showcase-my-story-at-viva-technology--148383.

Conclusion Part III

In the exploration of Chapters 21 to 30, we embarked on a journey through the intersections of blockchain technology and international business. This segment delved into the pivotal role of global payment systems, the facilitation of international trade through cryptocurrencies, the enhancement of supply chain visibility, and the challenges and solutions within cross-border contracts and regulatory compliance. Let's distill the insights gained into a comprehensive conclusion that illuminates the overarching themes.

The foundation was laid by highlighting the significance of global payment systems for international business. We discerned how blockchain, with its decentralized architecture, sets the stage for efficient cross-border transactions. Though programs like SWIFT and Ripple/XRP were briefly mentioned, subsequent chapters promise a detailed exploration. Then the focus shifted to the ways in which cryptocurrency serves as a catalyst for international trade. Beyond streamlining cross-border transactions, we explored how cryptocurrencies can underpin various facets of global commerce, fostering efficiency, transparency, and trust.

We next unraveled the pivotal role of supply chain visibility, showcasing how it amplifies trust in international markets. By leveraging blockchain's transparency, businesses can transform their supply chains, enhancing efficiency and reliability. The exploration of this topic contributes to a nuanced understanding of the intricacies of global business operations. We explored the challenges surrounding cross-border contracts and the importance of regulatory compliance. We comprehended how blockchain technology acts as a bulwark in mitigating these challenges. From smart contracts ensuring the execution of agreements to the imperative role of regulatory compliance, these chapters underscored the complexities and solutions within international business frameworks.

We extended our exploration by spotlighting the tokenization of global assets and the vital importance of identity management. We discerned how tokenization amplifies the efficiency of international transactions, while robust identity management on the blockchain ensures the security and legitimacy of participants. The next journey was the innovative aspects of international business. We examined how multi-lingual smart contracts overcome language barriers, fostering global collaborations. The emphasis on international partnerships elucidated the role of blockchain in creating synergies beyond borders.

The penultimate chapters underscored the significance of talent acquisition and cross-border marketing. As businesses expand globally, acquiring the right talent becomes imperative, and blockchain's role in facilitating cross-border marketing strategies was dissected. Collectively, these chapters weave an ecosystem of insights, showcasing the transformative potential of blockchain in navigating global frontiers. The blockchain emerges not merely as a technology but as a strategic enabler, revolutionizing the landscape of international business. Decentralization, security, transparency, and automation were recurrent themes, illustrating how these attributes align seamlessly with the demands of a globalized economy.

As we conclude Part III, it is evident that the terrain ahead holds promise and challenges in equal measure. The journey through these chapters has equipped us with an understanding of how blockchain technology, with its immutable ledger and decentralized architecture, serves as a lodestar in the ever-evolving realm of international business. The chapters ahead await, promising an even more profound exploration of the dance between blockchain and the global stage of commerce.

Part IV: the Web3 Wilderness: Unveiling the Mechanisms of Fintech

Introduction Part IV

As we set sail into the uncharted territories of Web3, Part IV beckons us to explore the digital wilderness that defines the next frontier of technological evolution. Chapters 31-40 of this odyssey unravel the intricacies of Web3, a realm where blockchain, decentralized finance (DeFi), artificial intelligence (AI), and cutting-edge technologies converge to redefine the landscape of modern business. Our journey commences with Chapter 31, where the Web3 lexicon is unveiled. This introduction to the language of the decentralized future lays the groundwork for understanding the nuanced terminologies that encapsulate the ethos of Web3. From smart contracts to consensus mechanisms, this lexicon becomes our guide through the maze of Web3's complex landscape.

In Chapter 32, we delve into the symbiotic relationship between blockchain and Web3. The decentralized ledger that once underpinned cryptocurrencies emerges as the cornerstone of decentralization. We explore how blockchain provides the structural integrity needed to fortify the decentralized ethos, ensuring transparency, security, and trust in the evolving digital paradigm. As we progress, Chapter 33 spotlights the pivotal role of artificial intelligence (AI) in Web3. We navigate through the transformative capabilities of AI, understanding how businesses that leverage its power gain a competitive edge. This chapter unveils AI as a driving force propelling us into a future where efficiency and adaptability are paramount.

Chapter 34 immerses us in the revolutionary tide of decentralized finance (DeFi). Beyond transforming financial transactions, DeFi emerges as a force decentralizing everything – challenging traditional notions of banking, lending, and asset management. Our exploration sheds light on the decentralized wave reshaping the financial landscape. Chapters 35 and 36 deepen our understanding of Web3 by exploring embedded finance and open banking. The former illuminates how financial services seamlessly integrate into daily activities, becoming intrinsic to various sectors. The latter unveils the collaborative potential of open banking within Web3's decentralized architecture, fostering innovation and interconnectedness.

In Chapters 37, 38, and 39, we navigate through niche domains within Web3, exploring the intersections of technology with insurance (InsureTech), regulation (RegTech), and conversational interfaces. Each chapter unveils how technology is revolutionizing these sectors, fostering innovation and reshaping traditional practices. Our odyssey reaches its zenith with Chapter 40, delving into the significance of digital wallets. Beyond being repositories for digital assets, these wallets emerge as custodians of identity in the Web3 domain. They encapsulate control, privacy, and security, embodying the essence of decentralized identity.

As we traverse through these chapters, we stand at the crossroads of an ever-evolving digital landscape. The interplay of blockchain, AI, DeFi, and emerging technologies paints a vibrant ecosystem that propels us toward a future where decentralized ideologies redefine how businesses operate. This introduction sets the stage for the revelations and insights awaiting us in the heart of the Web3 wilderness. As we embark on this odyssey, the chapters within Part IV invite us to witness the transformative potential of the digital frontier that lies ahead.

Chapter 31: Introduction to Web3, its terminology and its relation to business

Introduction

The evolution of the internet has reached a new frontier with Web3, reshaping how we perceive and engage with the digital realm. In this chapter, we embark on a journey, unraveling the intricacies of Web3 and deciphering the essential terminology that defines its landscape.[148] More importantly, we'll explore the profound implications of Web3 on modern business landscapes.

Web3 represents a paradigm shift from the traditional, centralized internet to a decentralized, user-centric model.[149] It leverages blockchain technology to enhance security, transparency, and user control. At its core, Web3 aims to redefine the way we interact with digital platforms, emphasizing user empowerment and privacy.

Terminology Deep Dive: Deciphering the Lexicon of Web3

Delving into the lexicon of Web3, we encounter terms such as decentralized applications (dApps), blockchain interoperability, tokenization, smart contracts, and decentralized autonomous organizations (DAOs). Each term plays a pivotal role in shaping the s of this new digital frontier.

- *Decentralized Applications (dApps):* These are applications built on blockchain technology, operating

---

[148] Layton, Ariana (2023). "Understanding Web3 Terminology from a non-technical perspective." Holder at https://www.holder.xyz/post/understanding-web3-terminology-from-a-non-technical-perspective.

[149] Staff (2023). "What Web3 Terminology do you need to know?" Meta Engine at https://www.metaengine.gg/blog/web3-gaming-terminology-and-definitions.

without a central authority. They epitomize the decentralized ethos of Web3, providing users with greater control over their data and interactions.[150]

- *Blockchain Interoperability:* As the Web3 ecosystem grows, the need for different blockchains to communicate and share information becomes crucial. Interoperability ensures seamless collaboration between diverse blockchain networks.[151]
- *Tokenization:* This process involves converting rights or assets into a digital token on a blockchain. Tokenization enables fractional ownership, facilitating new forms of value exchange and investment.[152]
- *Smart Contracts:* Self-executing contracts with predefined conditions directly written into code. They automate and enforce contract terms on the blockchain, streamlining processes across various industries.[153]
- *Decentralized Autonomous Organizations (DAOs):* These are entities governed by smart contracts, allowing for decentralized decision-making. DAOs represent a shift towards community-driven, democratic organizational structures.[154]

---

[150] Sharma, Rakesh (2023). "What is Decentralized Finance (DeFi) and How Does it Work?" Investopedia at https://www.investopedia.com/decentralized-finance-defi-5113835.

[151] Staff (2023). "What is Blockchain Interoperability?" Chainlink at https://chain.link/education-hub/blockchain-interoperability.

[152] Lutkevic, Ben (2023). "Tokenization" at https://www.techtarget.com/searchsecurity/definition/tokenization.

[153] Staff (2023). "What are smart contracts on the blockchain." IBM at https://www.ibm.com/topics/smart-contracts.

[154] Rieff, Nathan (2023). "Decentralized Autonomous Organizations (DAO): Definition, Purpose, and Example." Investopedia at https://www.investopedia.com/tech/what-dao/.

Web3 and Modern Business: A Symbiotic Relationship

Web3's impact on modern business is profound, offering transformative opportunities and posing unique challenges. As blockchain technology permeates various industries, businesses must adapt to this evolving landscape.

- *Decentralization and Trust:* Web3's decentralized nature enhances trust by eliminating reliance on central authorities. This is particularly relevant in sectors such as finance, where trust is paramount.[155]
- *Tokenization and Ownership:* The concept of tokenization facilitates new ownership models. Businesses can tokenize assets, allowing for more accessible investment opportunities and greater liquidity.[156]
- *Smart Contracts for Efficiency:* Smart contracts streamline processes, reducing reliance on intermediaries and automating complex transactions. This efficiency gain is especially valuable in supply chain management and legal processes.[157]

Conclusion

In conclusion, our exploration of Web3 and its terminology unveils a transformative landscape that is redefining the digital frontier. As modern businesses navigate this terrain,

---

[155] Yahya, Ali (2019). "Web 3.0 and the Future of Trust." A16zCrypto at https://a16zcrypto.com/posts/article/the-end-of-centralization-and-the-future-of-trust/. *See Contra* Nover, Scott (2022). "The Decentralized web is not decentralized." Quartz at https://qz.com/2112965/web3-is-not-decentralized.

[156] Staff (2023). "Unlocking Efficiency and Transparency: The Power of Smart Contracts." Ahmeddabad Computer Education at https://www.linkedin.com/pulse/unlocking-efficiency-transparency-power-smart/.

[157] *Id.*

understanding the nuances of Web3 becomes imperative. The symbiotic relationship between Web3 and modern business is a testament to the evolving nature of our digital interactions. As we chart the course forward, the promise of Web3 lies not just in its technology but in its potential to reshape the foundations of how we conduct business in the digital age.

Chapter 32: Blockchain in Web3: Foundation of Decentralization

Introduction

In the vast frontier of Web3, the relationship between blockchain and decentralization forms an unbreakable bond, reshaping the digital landscape. In this exploration, spanning 1700 words, we delve into the synergy between blockchain technology and Web3, unraveling how blockchain becomes the bedrock of decentralization. Decentralization is the defining ethos of Web3, a departure from traditional, centralized models. At its essence, decentralization redistributes control, fostering a more equitable and resilient digital ecosystem. Blockchain emerges as the enabler of this profound shift.

Blockchain's Role in Decentralization

Blockchain, a distributed ledger technology, underpins the decentralization narrative of Web3. Its core characteristics - immutability, transparency, and decentralization - form the backbone of a new era in digital interactions.

Immutability: Blockchain's ledger, once recorded, is unalterable. This immutability ensures a tamper-resistant record of transactions, establishing trust without the need for intermediaries.[158]

Transparency: Every participant in a blockchain network has access to a transparent, shared ledger. This transparency engenders trust and visibility, critical elements in the decentralized ethos.[159]

---

[158] Staff (2023). "What is immutable ledger in blockchain and its benefits?" Solu Lab at https://www.solulab.com/what-is-immutable-ledger-in-blockchain-and-its-benefits/.

[159] Hayes, *Supra*.

Decentralization: Operating on a peer-to-peer network, blockchain eliminates the need for central authorities. This decentralized architecture ensures that power is distributed among network participants.[160]

Smart Contracts: Pioneering Decentralized Automation

Smart contracts, executable code residing on the blockchain, epitomize the marriage of blockchain and decentralization. These self-executing contracts automate processes without intermediary intervention, facilitating trust in digital interactions.

Trustless Automation: The decentralized nature of smart contracts allows parties to engage in transactions without placing trust in a central authority. This trustless automation streamlines processes across industries.[161]

Immutable Execution: Once deployed on the blockchain, smart contracts execute exactly as coded, free from human interference. This immutability ensures the integrity of agreements, eliminating the risk of manipulation.[162]

Tokenization and Web3 Decentralization: Democratizing Ownership

Tokenization, another facet of blockchain technology, amplifies the decentralization journey. By converting real-world assets into digital tokens on the blockchain, tokenization democratizes ownership, offering fractional participation and liquidity.

---

[160] Staff (2024). "What is decentralization in blockchain?" AWS at https://aws.amazon.com/blockchain/decentralization-in-blockchain/.
[161] Staff (2023). "Blockchain and Automation: The Complete Guide." Supra at https://supraoracles.com/academy/blockchain-and-automation/.
[162] Solu Labs, *supra*.

Democratized Investment: Through tokenization, assets like real estate or art can be divided into tradable tokens. This opens up investment opportunities to a broader audience, challenging traditional exclusivity.[163]

Liquid Markets: Tokenized assets are traded on blockchain-based markets, fostering liquidity. This departure from traditional illiquid markets transforms the s of ownership and investment.[164]

Conclusion

In conclusion, the relationship between blockchain and Web3 is etching an enduring mark on the digital system. Blockchain, with its immutable ledger, transparent processes, and decentralized architecture, is not just a technological innovation; it's the cornerstone of a decentralized future. Smart contracts and tokenization, borne from blockchain's capabilities, amplify the ethos of trustless automation and democratized ownership.

As Web3 unfolds, the profound impact of blockchain in reshaping digital interactions becomes evident. It's a testament to the transformative power of technology when aligned with principles of decentralization. In this landscape, blockchain is not just a tool; it's a catalyst propelling us towards a decentralized era where trust, transparency, and ownership redefine how we engage with the digital world.

---

[163] Staff (2023). "Blockchain's Impact on Democratizing Investment." An Exploration of Invtron DAO." Invtron DAO at https://www.linkedin.com/pulse/blockchains-impact-democratizing-investment-exploration-invtron/.

[164] IG (2023). "Liquidity" at https://community.ig.com/forums/topic/23828-introduction-to-liquidity/. This is a good forum to go down the rabbit hole on different interpretations of liquidity.

Chapter 33: AI Technology Integration-Enhancing Fintech Operations

Introduction

Artificial Intelligence (AI): The Science of Making Machines that can think like humans.

In the vast expanse of Web3, the emergence of Artificial Intelligence (AI) stands as a beacon, guiding businesses through uncharted territories. In this exploration, we delve into the pivotal role of AI in the Web3 landscape, unraveling its transformative impact, the consequences for companies lagging behind, and how it serves as an indispensable tool streamlining business operations.

Artificial Intelligence, once a promise of the future, is now the driving force shaping the contours of Web3.[165] Its ability to mimic human intelligence and adapt in real-time positions it as a cornerstone technology in the evolving digital era. In Web3, companies not harnessing the power of AI risk fading into obsolescence. The ability of AI to analyze vast datasets, extract meaningful insights, and predict trends offers a competitive advantage. From personalized user experiences to data-driven decision-making, AI is a catalyst for business evolution.[166]

Transformative Impact on Industries: A Glimpse into Tomorrow's Business Landscape

Finance: AI algorithms are revolutionizing financial services, enhancing fraud detection, automating trading strategies, and personalizing investment advice. The integration of AI in

---

[165] Pattam, Aruna (2024). "Artificial Intelligence, defined in simple terms." HCL Tech at https://www.hcltech.com/blogs/artificial-intelligence-defined-simple-terms.

[166] *Id.*

decentralized finance (DeFi) platforms is reshaping how financial transactions occur.[167]

- *Healthcare:* In the healthcare sector, AI-driven diagnostics, predictive analytics, and personalized medicine are advancing patient care. Blockchain-AI integration ensures secure, transparent health data management.[168]
- *Supply Chain:* AI optimizes supply chain processes, ensuring efficiency and reducing costs. Smart contracts on the blockchain, fueled by AI insights, automate, and enhance supply chain visibility, fostering transparency and traceability.[169]

The Consequences of Lagging Behind: A Digital Chasm

Companies resistant to AI integration risk falling into a digital chasm. In Web3, where agility and innovation are paramount, failure to leverage AI means missing opportunities, succumbing to inefficiencies, and losing the competitive edge.

- *Inefficiencies and Inertia:* AI streamlines operations, automates repetitive tasks, and enhances productivity. Companies resisting this transformative technology risk

---

[167] Staff (2024). "What is Artificial Intelligence in Finance." Google Cloud at https://cloud.google.com/discover/finance-ai.

[168] Davenport, Thomas (2019). "The potential for Artificial Intelligence in Healthcare." NIH at https://www.ncbi.nlm.nih.gov/pmc/articles/PMC6616181/.

[169] Staff (2023). "AI in Supply Chain." Kinaxis at https://www.kinaxis.com/en/ai-supply-chain?utm_source=google&utm_medium=ppc&utm_campaign=7015Y000004C4aeQAC&utm_term=155403930924&utm_content=&gad_source=1&gclid=CjwKCAiAkp6tBhB5EiwANTCx1BupUFCOw6MVhBtyyeO9nSY1rFtxaR5cqnMuq-pOU5_u6dhO_-b_UBoCSbgQAvD_BwE.

being burdened by inefficiencies and inertia in a landscape where adaptability is key.[170]

- *Lost Opportunities:* AI uncovers opportunities hidden within data, guiding strategic decision-making. Businesses not leveraging AI may miss out on discovering untapped markets, predicting consumer trends, and optimizing resource allocation.[171]

AI as a Business Ally: Streamlining Operations, Enhancing Efficiency

When looking at AI as a business ally, you have the ability to have workers doing work even when your workers are not there. Automation allows for systems to analyze numbers, oversee productions, and manage the supply chain while your workers are exploring a better quality of life.

- *Automated Operations:* AI automates mundane tasks, freeing up human resources for strategic, creative endeavors. From customer service chatbots to automated data analysis, AI simplifies operations.[172]
- *Data-Driven Decision-Making:* AI sifts through vast datasets, extracting actionable insights. In Web3, where data reigns supreme, AI becomes the compass guiding businesses toward informed, data-driven decisions.[173]

---

[170] Buchanan, Naomi (2023). "Companies waiting to integrate AI risk being left behind." Investopedia at https://www.investopedia.com/companies-waiting-to-integrate-ai-risk-being-left-behind-report-finds-7970970.

[171] *Id.*

[172] Staff (2023). "What is AI Automation? Explore Applied Uses and Productivity." UpWork at https://www.upwork.com/resources/ai-automation.

[173] Sahota, Neil (2023). "Leveraging AI for Data-Driven-Decision Making While Safeguarding Privacy and Security." Forbes at https://www.forbes.com/sites/neilsahota/2023/12/06/leveraging-ai-

Conclusion

In summary, AI stands to guide businesses through the uncharted territories of Web3. Its transformative impact spans industries, offering a competitive advantage and shaping tomorrow's business landscape. The consequences of lagging behind are profound, leading to inefficiencies and missed opportunities.

AI is not just a technology; it's a business ally streamlining operations, enhancing efficiency, and facilitating data-driven decision-making. In the landscape of Web3, where adaptability is the currency of success, AI emerges as an indispensable tool, a beacon illuminating the path toward a future where businesses thrive on innovation, agility, and the seamless integration of artificial intelligence.

---

for-data-driven-decision-making-while-safeguarding-privacy-and-security/?sh=10c914617a0f.

Chapter 34: Decentralized Finance (DeFi)-Decentralizing Everything.

Introduction

As the digital frontier expands in Web3, Decentralized Finance (DeFi) emerges as a revolutionary force reshaping the financial landscape. In this exploration spanning 2000 words, we embark on a deep dive into DeFi, dissecting its core principles, understanding its transformative impact, and unveiling how decentralization is becoming the catalyst for democratizing financial systems worldwide.

DeFi Principles: Unraveling the Fabric of Decentralization

At the heart of DeFi lies the fundamental principle of decentralization. Unlike traditional finance, which relies on centralized authorities like banks and governments, DeFi operates on blockchain networks, ensuring transparency, security, and accessibility.

- *Smart Contracts and Automated Protocols:* DeFi platforms utilize smart contracts, self-executing code on the blockchain, to automate financial processes. Automated protocols govern lending, borrowing, trading, and yield farming, eliminating the need for intermediaries.[174]
- *Open Access and Financial Inclusion:* DeFi opens the gates of financial services to anyone with an internet connection, fostering financial inclusion. Individuals worldwide can access and participate in DeFi platforms,

---

[174] Staff (2022). "Open Finance v. Decentralized Finance" Cryptopedia at https://www.gemini.com/cryptopedia/open-vs-decentralized-finance-defi.

transcending geographical barriers and traditional banking limitations.[175]

Transformative Impact: Decentralizing Financial Systems

DeFi platforms facilitate global participation, enabling individuals from underserved regions to access financial services. The unbanked and underbanked gain entry to lending, trading, and investment opportunities.

DeFi eliminates the need for intermediaries like banks, reducing dependence on centralized authorities. This disintermediation leads to faster and more cost-effective financial transactions. As the digital space becomes more competitive, each unit of currency cut from the cost could be crucial in ensuring the success of a business.

The DeFi Ecosystem: A Plethora of Possibilities

- *Lending and Borrowing Protocols:* DeFi introduces decentralized lending and borrowing protocols where users can lend their assets to earn interest or borrow assets against collateral without relying on traditional banks.
- *Decentralized Exchanges (DEXs):* DEXs operate without intermediaries, allowing users to trade directly from their wallets. This enhances security, privacy, and eliminates the risk of exchange hacks.[176]
- *Yield Farming and Liquidity Mining:* DeFi platforms incentivize users through yield farming and liquidity

---

[175] *Id.*

[176] Staff (2024). "What is a Dex?" Coinbase at https://www.coinbase.com/learn/crypto-basics/what-is-a-dex.

mining, where participants earn rewards for providing liquidity to decentralized liquidity pools.[177]

Challenges and Risks: Navigating the DeFi Wilderness

Despite the potential, DeFi is not without challenges. Smart contract vulnerabilities and exploits pose risks to users' funds, emphasizing the need for robust security measures and auditing.[178]

The decentralized nature of DeFi presents regulatory challenges. Regulatory uncertainty, compliance issues, and potential interventions from authorities are aspects that DeFi platforms must navigate.[179]

Conclusion

In conclusion, DeFi marks a paradigm shift in the financial landscape, decentralizing everything from lending and borrowing to trading and yield farming. Its core principles of transparency, accessibility, and disintermediation foster a global financial ecosystem that transcends traditional constraints.

DeFi's transformative impact lies in its ability to democratize finance, providing global access, reducing dependence on intermediaries, and introducing a plethora of decentralized financial services. Challenges and risks exist, but the decentralized ethos continues to drive innovation and resilience in the DeFi wilderness.

---

[177] Staff (2023). "Understanding yield farming and liquidity mining." Reintech at https://reintech.io/blog/understanding-yield-farming-liquidity-mining.

[178] Staff (2023). "Dangers of Smart Contracts." Contract Logix at https://www.contractlogix.com/contract-management/dangers-of-smart-contracts/.

[179] Smithmyer, Christopher; Alli, Remi; Jadoon, AJ & Ashe, Ryan (2022). Dragons of the Digital Age. Elite Exclusivity: Detroit.

As we navigate the uncharted territories of Web3, DeFi stands as a testament to the evolving nature of finance, where decentralization becomes the catalyst for financial inclusion, innovation, and a reimagined global economy.

Chapter 35: Embedded Finance- Integrating Financial Services in Apps and Webpages

Introduction

Embedded Finance is a force of nature in the Web3 wilderness, reshaping the traditional paradigms of financial services. In this exploration, we delve into the core of embedded finance, unraveling its principles, understanding its impact on international business, and forecasting its role in the ever-evolving landscape of the digital economy.

Embedded Finance Unveiled: The Integration of Financial Services

Embedded finance represents the integration of financial services into non-financial platforms seamlessly. Instead of relying on standalone financial institutions, businesses embed financial functionalities directly into their products or services.[180] This integration is made possible by leveraging Application Programming Interfaces (APIs) and advanced technologies.

Embedded finance operates in alignment with the decentralized ethos of Web3. It breaks down traditional financial silos, allowing businesses to offer financial services without the need for traditional banking intermediaries.[181]

Impact on International Business: Reshaping the Global Economic Landscape

One of the primary impacts of embedded finance on international business lies in its ability to streamline cross-

---

[180] Sullivan, Tom (2022). "What is embedded Finance? 4 ways it will change fintech." Plaid at https://plaid.com/resources/fintech/what-is-embedded-finance/.

[181] *Id.*

border transactions.[182] By embedding payment functionalities directly into their platforms, businesses can facilitate seamless and cost-effective international transactions.

Embedded finance contributes to financial inclusion on a global scale.[183] By integrating financial services into various products, businesses can extend their reach to underserved regions, providing access to banking, lending, and investment opportunities.

The Role of Blockchain in Embedded Finance: Ensuring Security and Transparency

The decentralized nature of blockchain technology enhances the security of embedded finance.[184] Transactions conducted through embedded financial services can leverage blockchain's immutability and cryptographic security, reducing the risk of fraud and ensuring the integrity of financial data.

Blockchain provides a transparent and auditable ledger for financial transactions. This transparency builds trust among users, crucial for international business operations where trust is paramount.[185]

---

[182] Staff (2023). "Embedded Finance: a Global POV." Workweek at https://workweek.com/2023/09/05/embedded-finance-a-global-pov/.

[183] Staff (2023). "Unleashing the potential of embedded finance: Revolutionizing convenience and trade." Global Trade Review at https://www.gtreview.com/news/fintech/unleashing-the-potential-of-embedded-finance-revolutionising-convenience-and-trade/.

[184] Staff (2023). "The Rise of Embedded Finance with Blockchain." Gyan Consulting at https://www.linkedin.com/pulse/rise-embedded-finance-blockchain-gyanconsulting/.

[185] Staff (2023). "Embedded Finance: What is it and how can it change traditional banking?" Sydle at https://www.sydle.com/blog/embedded-finance-64a70d850efb3b42f7522130.

Navigating the Embedded Finance Landscape

As embedded finance expands its footprint, navigating regulatory landscapes becomes paramount. Businesses must ensure compliance with financial regulations in various jurisdictions, presenting a challenge in the complex world of international finance.

The integration of financial services into non-financial platforms raises security concerns. Businesses must implement robust cybersecurity measures to safeguard user data and financial transactions from potential threats.

Conclusion

In conclusion, embedded finance emerges as a transformative force, paving the way for the future of finance in the international business community. Its seamless integration of financial services into non-financial platforms aligns with the decentralized principles of Web3, offering new possibilities for cross-border transactions, financial inclusion, and global accessibility.

The role of blockchain in ensuring security and transparency underscores the potential of embedded finance to foster trust in international business operations. However, challenges such as regulatory compliance and security considerations require careful navigation.

As businesses continue to embed financial functionalities into their products and services, the global economic landscape undergoes a paradigm shift. Embedded finance becomes not only a facilitator of international transactions but a catalyst for financial empowerment and inclusivity in the interconnected world of Web3.

Chapter 36: Open Banking- Facilitating Collaboration in Financial Services

Introduction

In the vast expanse of Web3, the concept of open banking emerges as a catalyst for reshaping the financial ecosystem. With its roots deeply intertwined with the principles of decentralization and interconnectedness, open banking stands at the forefront of innovation. In this exploration spanning 1600 words, we delve into the essence of open banking, unraveling its intricacies, understanding its relationship with Web3, and envisioning its transformative impact on the financial landscape.

A Paradigm of Accessibility and Collaboration

> *Open Banking: systems that allow secure information access to financial data through the use of application programming interfaces (APIs)*

Open banking is a financial model that advocates for the use of open Application Programming Interfaces (APIs) to enable third-party developers to build applications and services around financial institutions. Instead of a closed and proprietary system, open banking promotes transparency, collaboration, and accessibility. At the heart of Web3's ethos, open banking resonates with the decentralized paradigm. It breaks down the traditional silos of financial institutions, allowing for a more interconnected and user-centric financial experience.

- *Blockchain and Smart Contracts:* Web3, with its foundation in blockchain technology, provides a secure and transparent infrastructure for open banking.[186]

---

[186] Kazoulis, Antonis (2023). "How does open banking shape the future of crypto KYC." Go Cardless at

Smart contracts, executed on decentralized networks, facilitate automated and trustless transactions, enhancing the efficiency of open banking protocols.

- *Decentralized Identity (DID):* In the Web3 wilderness, decentralized identity solutions become integral to open banking.[187] DID, leveraging blockchain for secure and verifiable identity management, aligns with the principles of user-centric control and privacy, crucial for the open banking ecosystem.
- *Enhanced Access to Banking Services:* Open banking, in synergy with Web3, has the potential to enhance access to banking services globally. By fostering collaboration between financial institutions and third-party developers, it opens avenues for innovative financial products and services, catering to diverse user needs.[188]
- *Financial Inclusion for the Unbanked:* The decentralized nature of Web3, coupled with open banking principles, can address the challenge of financial inclusion.[189] By leveraging blockchain's global reach and open APIs, open banking can extend its benefits to the unbanked and underserved populations.

---

https://gocardless.com/guides/posts/open-banking-future-kyc-crypto/.

187 Staff (2022). "Open Banking: Definition, How it Works, and Risks." Investopedia at https://www.investopedia.com/terms/o/open-banking.asp.

188 Staff (2023). "Open Banking Explained: What it is and how its changing financial services." Stripe at https://stripe.com/resources/more/open-banking-explained.

189 Staff (2023). "Financial Inclusion via open banking around the world." Mastercard at https://www.mastercardservices.com/en/advisors/archived-practices/open-banking/insights/purposeful-profitable-financial-inclusion-open-0.

Challenges and Considerations: Navigating the Open Banking Frontier

The interconnected nature of open banking raises concerns about data security and privacy. Web3's emphasis on cryptographic principles can provide robust solutions, yet challenges persist in ensuring secure data sharing.

As open banking disrupts traditional financial models, navigating regulatory landscapes becomes critical. The need for harmonized regulations that foster innovation while ensuring consumer protection remains a challenge.

Conclusion

In summary, open banking emerges as a beacon of financial evolution in the Web3 landscape. Its principles of transparency, collaboration, and accessibility align seamlessly with the decentralized ethos of Web3. As blockchain and smart contracts become integral to open banking protocols, the stage is set for a transformative journey in the financial ecosystem.

The impact on financial inclusion, with the potential to democratize access to banking services, highlights the societal benefits of open banking in the Web3 era. Challenges such as data security and regulatory considerations beckon for thoughtful navigation. Yet, as the open banking frontier unfolds in the vast expanse of Web3, it paves the way for a future where financial services are not just open but woven into the fabric of decentralized and interconnected possibilities.

Chapter 37: Insurtech- Technological Innovation in Insurance

Introduction

In the expansive realm of Web3, innovation continues to reshape industries, and insurance is no exception. Insurtech, the convergence of insurance and technology, takes center stage as a transformative force. This exploration delves into the intricacies of Insurtech within the Web3 paradigm. We unravel how the principles of decentralization, blockchain, and smart contracts are influencing the evolution of the insurance industry.

Insurtech Unveiled: Reinventing Insurance in the Web3 Era

Insurtech, a portmanteau of insurance and technology, represents a seismic shift in the insurance landscape. At its core, it leverages technological advancements to enhance and streamline the entire insurance value chain – from underwriting and pricing to claims processing and customer experience.[190]

The principles of Web3 align seamlessly with the transformative goals of Insurtech. Decentralization, a cornerstone of Web3, fosters transparency and trust in insurance processes.[191] Blockchain, the distributed ledger technology, serves as the bedrock for secure and transparent data management, addressing longstanding issues of information asymmetry in the insurance sector.

---

[190] Ricard, Paul & Zwick, Joshua (2023). "Will Web3 Reinvest Insurance?" Marsh Mclennan at https://www.marshmclennan.com/insights/publications/2022/october/will-web3-reinvent-insurance.html.

[191] *Id.*

Smart Contracts: The Building Blocks of Insurtech in Web3

Smart contracts, epitomizing the self-executing and programmable nature of Web3, play a pivotal role in reshaping insurance processes.[192] These digital contracts, deployed on blockchain networks, automate and enforce contractual agreements without the need for intermediaries. In the realm of Insurtech, smart contracts revolutionize claims processing, policy issuance, and even parametric insurance solutions.

Parametric insurance, a cutting-edge Insurtech solution, utilizes smart contracts to trigger payouts based on predefined parameters. In the Web3 era, this synergy eliminates bureaucratic delays and enhances the speed and accuracy of claims settlements. Decentralized and automated, parametric insurance aligns with the ethos of Web3 – efficiency, transparency, and user-centricity.[193]

Decentralized Identity (DID): Transforming KYC in Insurance

Web3 introduces the concept of Decentralized Identity (DID), a revolutionary solution for Know Your Customer (KYC) processes. In the Insurtech landscape, DID can streamline onboarding procedures, enhance data security, and provide users with control over their personal information.

DID empowers insurance customers by allowing them to control access to their identity information. This user-centric approach

---

[192] Staff (2023). "The future of smart contracts in insurance." Norton Rose Fulbright at https://www.nortonrosefulbright.com/en/knowledge/publications/88244592/the-future-of-smart-contracts-in-insurance.

[193] *Id.*

not only enhances privacy but also reduces the risk of data breaches, a prevalent concern in the insurance sector.[194]

While the marriage of Insurtech and Web3 promises transformative benefits, challenges and considerations are inherent. Privacy concerns, regulatory frameworks, and interoperability of blockchain platforms require thoughtful navigation. As the Insurtech frontier unfolds within Web3, industry stakeholders must collectively address these challenges to unlock the full potential of innovation.

Conclusion

In conclusion, Insurtech's journey within the Web3 landscape symbolizes a paradigm shift in the insurance industry. The principles of decentralization, smart contracts, and decentralized identity redefine traditional insurance processes, paving the way for greater efficiency, transparency, and user empowerment.

As the industry grapples with challenges, the transformative potential of Insurtech in Web3 remains undeniable. The synergy between technological innovation and decentralized principles promises a future where insurance becomes more accessible, user-centric, and resilient. Insurtech's Web3 odyssey is not just an evolution; it is a testament to the ongoing digital revolution reshaping the fundamental fabric of the insurance sector.

---

[194] Davis, Christine (2021). "Insuretech Concepts: The Internet of Things." ABA at https://www.americanbar.org/groups/tort_trial_insurance_practice/publications/tortsource/2021/spring/insurtech-concepts-internet-things/.

Chapter 38: Reprise of RegTech – Regulatory Technology for Compliance and the Constant Need for Evolution.

Introduction

The intersection of Regulatory Technology (RegTech) and Web3 marks a transformative juncture for businesses navigating the complexities of compliance in the digital age. This in-depth analysis explores the profound impact of RegTech within the context of Web3 integration. As businesses embrace decentralized technologies, the role of RegTech becomes pivotal in fostering regulatory compliance, risk management, and operational resilience.

RegTech Unveiled: Navigating the Regulatory Landscape in Web3

RegTech represents the fusion of technology and regulatory processes to streamline compliance activities efficiently. In the context of Web3, where decentralized principles underscore innovation, RegTech emerges as a facilitator of seamless integration between regulatory requirements and decentralized business models.[195]

To make life easier for everyone, insuretech automates the process. Web3 technologies, particularly blockchain and smart contracts, provide a fertile ground for automating compliance processes. RegTech solutions leverage the transparency and immutability of blockchain to ensure that compliance requirements are met without compromising the decentralized ethos.

---

[195] Staff (2023) "Fintech's new wave of innovation fueled by Web3." Maxcode at https://www.maxcode.net/blog/fintechs-new-wave-of-innovation-fueled-by-web-3-0/.

The Synergy of Decentralization and Compliance in Web3

Web3, characterized by decentralization, aligns with the goals of RegTech in reshaping compliance frameworks. The decentralized nature of blockchain ensures that compliance data is securely stored, auditable, and resistant to tampering.[196] Smart contracts, as self-executing agreements, automate compliance tasks, reducing the burden on businesses and enhancing operational efficiency.[197]

One of the of the essential elements of RegTech in the Web3 era is the integration of Decentralized Identity (DID). By utilizing decentralized identifiers and verifiable credentials, businesses can streamline Know Your Customer (KYC) processes, providing a secure and user-centric approach to identity verification.[198]

Risk Management in a Decentralized Landscape

Web3 introduces new paradigms of risk in decentralized ecosystems. RegTech plays a crucial role in developing risk management frameworks that align with the and rapidly evolving nature of decentralized technologies.

As businesses explore tokenomics within the Web3 landscape, RegTech solutions become instrumental in ensuring compliance

---

[196] FM (2023). "Regtech in the era of decentralized finance (DeFi): Addressing Compliance Challenges." Finance Magnates at https://www.financemagnates.com/institutional-forex/regulation/regtech-in-the-era-of-decentralized-finance-defi-addressing-compliance-challenges/.

[197] Staff (2023). "Decentralization of Regtech and Building Trust in Big Banks." Planet Compliance at https://www.planetcompliance.com/decentralization-of-regtech/.

[198] Staff (2023). "How Regtech can automate AML and KYC Compliance." Sanction Scanner at https://sanctionscanner.com/blog/how-regtech-can-automate-kyc-and-aml-compliance-179.

with regulatory frameworks governing token issuance, trading, and utilization.[199] The transparent and auditable nature of blockchain, coupled with RegTech protocols, facilitates regulatory adherence in tokenized ecosystems.

Challenges and Considerations in Web3-RegTech Integration

While the marriage of Web3 and RegTech offers transformative possibilities, challenges and considerations persist. Interoperability between diverse blockchain networks, evolving regulatory landscapes, and the need for standardization pose hurdles that demand nuanced solutions.

The decentralized nature of Web3 introduces a plethora of blockchain networks. RegTech solutions must navigate the challenges of interoperability and standardization to ensure consistent compliance processes across diverse ecosystems.

Conclusion

In conclusion, the integration of RegTech with Web3 signifies a pivotal advancement in how businesses approach regulatory compliance. The synergy between decentralized technologies and RegTech solutions holds the promise of creating more resilient, efficient, and transparent compliance ecosystems.

As businesses embark on the Web3 journey, the role of RegTech becomes not just a necessity but a strategic enabler. The symbiosis of decentralization and regulatory compliance propels industries toward a future where innovation thrives within the boundaries of legal and regulatory frameworks, fostering a harmonious coexistence in the digital landscape.

---

[199] *Id.*

Chapter 39: Conversational Banking-Humanizing Financial Interactions

Introduction

As the Web3 landscape unfolds, the concept of conversational banking emerges as a bridge between technological innovation and the indispensable human touch in financial interactions. This chapter analyzes the significance of conversational banking within the decentralized cryptoverse, emphasizing the role of human-centric communication in a realm dominated by blockchain, smart contracts, and decentralized finance (DeFi).

The Evolution of Conversational Banking in Web3

Conversational banking refers to the integration of natural language processing (NLP), artificial intelligence (AI), and blockchain technologies to facilitate seamless, human-like interactions between users and decentralized financial services.[200] In the Web3 era, where decentralization is paramount, conversational banking acts as a user-friendly interface, making the complex world of blockchain accessible to a broader audience.

Web3 technologies, while revolutionary, can be daunting for users unfamiliar with the intricacies of blockchain and decentralized applications. Conversational banking acts as a humanizing layer, enabling users to interact with the cryptoverse using familiar language and conversational interfaces.

The Role of AI and NLP in Conversational Banking

In a decentralized financial ecosystem dominated by smart contracts and blockchain protocols, conversational banking leverages AI and NLP to enhance the user experience. Through

---

[200] Staff (2024). "Conversational Banking: The Complete Guide for 2024." Sinch at https://www.sinch.com/blog/conversational-banking.

intuitive chatbots and voice-activated interfaces, users can interact with their decentralized financial assets, execute transactions, and seek information using natural language.[201]

AI-powered conversational banking systems can provide personalized financial assistance by analyzing user preferences, transaction history, and market trends. This level of personalization enhances the user's ability to make informed decisions within the decentralized financial landscape.

Navigating Transactions and Inquiries in Web3

Conversational banking streamlines transaction execution in the Web3 cryptoverse. Users can initiate and confirm transactions through simple conversational prompts, reducing the complexity associated with interacting directly with smart contracts. [202]Through natural language queries, users can retrieve real-time information about their decentralized assets, market trends, and portfolio performance. Conversational interfaces serve as a conduit for users to stay informed and make data-driven decisions.

The Human Element in the Decentralized Future

Despite the automated and decentralized nature of Web3, the human element remains essential. Conversational banking introduces a layer of trust and assurance by providing users with a familiar and empathetic interface. Users can seek guidance, clarify doubts, and receive support through human-like interactions, fostering a sense of security in the cryptoverse. Conversational banking contributes to financial inclusion within the Web3 ecosystem.[203] By simplifying interactions and

---

[201] McGloin, Cathal (2023). "Conversational AI in Banking: 6 Engaging Examples. Servisbot at https://servisbot.com/conversational-ai-in-banking/.

[202] *Id.*

[203] Sinch (2024), *Supra*.

eliminating the need for specialized technical knowledge, conversational interfaces open the doors of decentralized finance to a broader demographic, fostering inclusivity.

Conclusion

In conclusion, conversational banking stands as a testament to the harmonious fusion of technological innovation and the human touch in the Web3 cryptoverse. As decentralized finance continues to reshape the financial landscape, conversational interfaces serve as a gateway, ensuring that users, regardless of their technical proficiency, can seamlessly navigate and engage with the decentralized future.

In the evolving narrative of Web3, where decentralization and automation reign, conversational banking remains a testament to the enduring importance of human-centric design and communication. The cryptoverse, guided by the principles of inclusivity and accessibility, welcomes users into a future where the complexities of blockchain are navigable through the simplicity of conversation.

Chapter 40: Digital Wallets- Secure and Convenient Asset Management.

Introduction

As the Web3 era unfolds, digital wallets emerge as the bulwark connecting individuals to the decentralized fabric of the internet. This chapter embarks on an exploration of digital wallets, elucidating their profound significance, pivotal role in the Web3 ecosystem, and the transformative concept of digital identity they embody.

Defining Digital Wallets in the Web3 Realm

A digital wallet in the Web3 context is not merely a repository for storing cryptocurrency; it embodies the essence of decentralization. It is a secure, user-centric interface that empowers individuals with control over their digital assets, transactions, and, most crucially, their identity in the decentralized landscape.[204]

Beyond cryptocurrencies, modern digital wallets serve as custodians for a spectrum of digital assets, including non-fungible tokens (NFTs), decentralized finance (DeFi) tokens, and various other blockchain-based assets. This expansion underscores their evolving role as comprehensive repositories in the Web3 ecosystem.

The Importance of Digital Wallets in Web3

Digital wallets function as gateways, facilitating entry into the expansive realm of decentralized applications (DApps), smart contracts, and blockchain-based services. They embody a user's sovereign connection to the decentralized internet, allowing seamless interaction with various decentralized protocols.

---

[204] Staff (2023). "A Complete Overview of Web3 Wallets." Quicknode at https://www.quicknode.com/guides/web3-fundamentals-security/basics-to-web3-wallets.

In the Web3 landscape, digital wallets serve as the embodiment of user control and ownership. Users hold the cryptographic keys within their wallets, providing unprecedented control over their assets. This control aligns with the fundamental ethos of decentralization, liberating individuals from traditional intermediaries and centralized control.

Digital Wallets as the Cornerstone of Web3 Identity

> *Digital Wallets: a type of financial app which gives a person access to their financial information online; in blockchain, these also are the key to your digital signature.*

Web3 introduces a paradigm shift where identity is not confined to physical documents but encoded in cryptographic keys. Digital wallets, through their association with private keys, become the digital embodiment of one's identity in the decentralized space. They signify the fusion of the physical and digital realms, where ownership and identity are cryptographically secured.[205]

Digital wallets enhance privacy by enabling pseudonymous interactions. Users can engage with decentralized applications and conduct transactions without revealing their real-world identities.[206] This pseudonymity fosters a more private and secure online experience, aligning with the principles of user empowerment and data protection.

Securing Transactions and Interactions

Digital wallets leverage cryptographic algorithms to secure transactions and interactions. Private keys, generated within the

---

[205] Community (2024). "Digital ID Wallets Explained." Cloudentity Portal at https://cloudentity.com/developers/basics/identity/id-wallets/.

[206] *Id.*

confines of the wallet, serve as the sentinel guarding access to assets and ensuring the integrity of interactions.[207] This cryptographic foundation reinforces the security of Web3 identity.

Beyond transactions, digital wallets contribute to the development of a web of trust and reputation in the decentralized space. As users engage with various platforms, their digital wallets accrue a history of interactions, building a reputation that adds a layer of trust to decentralized exchanges, collaborations, and participatory endeavors.

Conclusion

In conclusion, digital wallets in the Web3 landscape unravel a digital identity, weaving together control, privacy, and security. Beyond serving as repositories for assets, they encapsulate the spirit of decentralization, offering users a sovereign gateway to the expansive possibilities of the decentralized internet.

The importance of digital wallets extends beyond their technical functionalities; they symbolize the democratization of identity in the digital age. As custodians of cryptographic keys, digital wallets empower individuals with unparalleled control over their digital assets and interactions. In the evolving narrative of Web3, digital wallets emerge not just as tools but as integral components shaping the very fabric of decentralized identity and ownership.

---

[207] Staff (2024). "What is a digital wallet and is it safe?" Citizens Bank at https://www.citizensbank.com/learning/is-digital-wallet-safe.aspx.

Conclusion Part IV

As we conclude the exploration of Part IV, encompassing Chapters 31-40, we emerge from the depths of the Web3 wilderness with a newfound understanding of its terrain. This digital odyssey has been a journey through the pillars of Web3, unraveling the ecosystem that interlaces blockchain, decentralized finance (DeFi), artificial intelligence (AI), and the foundational elements shaping the decentralized future.

In delving into Web3's lexicon in Chapter 31, we deciphered the complex terminology that forms the bedrock of this transformative era. From decentralized applications (DApps) to consensus mechanisms, the lexicon is the language through which we articulate the principles reshaping modern business landscapes. Chapter 32 highlighted the symbiotic relationship between blockchain and Web3, underscoring how the former serves as the foundation of decentralization. Blockchain, once the underpinning of cryptocurrencies, emerges as the backbone that fortifies the decentralized ethos, fostering transparency, security, and trust.

The role of AI in Chapter 33 was spotlighted as a bulwark in the Web3 narrative. As a driving force, AI propels businesses into the future, bestowing them with the agility and efficiency needed to navigate the evolving digital landscape. Those not harnessing the power of AI risk being left behind in the wake of rapid technological advancement. Chapter 34 ushered us into the realm of decentralized finance (DeFi), where a tidal wave of decentralization is sweeping across traditional financial paradigms. This chapter unveiled how DeFi isn't just transforming financial transactions; it's decentralizing everything, challenging traditional notions of banking, lending, and asset management.

Chapters 35 and 36 took us deeper into the Web3 landscape, exploring embedded finance and open banking. The former highlighted how financial services seamlessly integrate into everyday activities, while the latter illuminated the collaborative potential of open banking within Web3's decentralized architecture.

Chapters 37, 38, and 39 delved into specific niches within Web3, exploring the intersection of technology with insurance (InsureTech), regulation (RegTech), and conversational interfaces. Each segment showcased how Web3 is not a monolithic entity but a diverse ecosystem fostering innovation across industries. The journey reached its pinnacle with Chapter 40, elucidating the pivotal role of digital wallets. Beyond being repositories for digital assets, digital wallets emerged as the custodians of identities in the Web3 domain. They encapsulate control, privacy, and security, embodying the essence of decentralized identity.

In concluding Part IV, we stand at the cusp of a future landscape shaped by the foundational elements explored in these chapters. The interconnectedness of blockchain, AI, decentralized finance, and emerging technologies paints a vibrant picture of a decentralized world where control, transparency, and trust redefine how businesses operate.

This digital odyssey through the Web3 wilderness serves as a prelude to the uncharted territories awaiting exploration in the subsequent parts of our journey. As we navigate the unfolding narrative of Web3, the chapters within Part IV lay the groundwork for understanding the symbiotic relationship between technology and decentralized ideologies, propelling us toward a future where the digital and physical realms seamlessly converge.

Part V: Cross-Border Exploration- The Frontier of Payments

Introduction to Part V

In the narrative of the cryptoverse, the exploration of cross-border payments within the Web3 paradigm stands as a pivotal focus. As we embark on Part V, delving into the innovations that define the crossroads of decentralized technologies and global financial interactions, we find ourselves at the frontier of a transformative era.

Chapter 41 - Introduction to Cross-Border Payments in Web3: Our journey begins with a fundamental exploration of cross-border payments within the expansive landscape of Web3. This chapter lays the groundwork, elucidating the fundamental principles that underpin international transactions in this decentralized wilderness. Chapters 42-43 - Unveiling SWIFT's Legacy and SWIFT 2.0: The subsequent chapters unfold the historical ecosystem of SWIFT, the traditional backbone of international transactions, and its evolution into SWIFT 2.0. Examining these mechanisms sets the stage for understanding the challenges that Web3 poses to established systems.

Chapter 44 - Ripple/XRP: Navigating Efficiency in Cross-Border Transactions: A deep dive into the Ripple/XRP ecosystem sheds light on an alternative approach to cross-border payments. This chapter illuminates the areas where Ripple/XRP excels and where it encounters challenges, setting the tone for the evolving s in this domain. Chapter 45 - The Self-Validating Nature of Cryptocurrency: As we move forward, the self-validating attributes of cryptocurrency take center stage. This chapter explores how the intrinsic properties of digital currencies can outshine traditional systems like SWIFT, especially in the context of international payments.

Chapter 46 - Jurisdictional Challenges in Cross-Border Payments: Navigating the global landscape, we confront the jurisdictional challenges that arise when decentralized technologies intersect with traditional legal frameworks. Understanding these challenges becomes crucial as we envision a future where borders become more fluid in the digital realm. Chapter 47 - The Call for a Unified Regulatory System: Recognizing the need for harmonized regulations becomes imperative. This chapter advocates for a unified regulatory system to address the complexities arising from decentralized technologies, fostering an environment of clarity and confidence.

Chapter 48 - Settlement Functionality and Tractability: The importance of settlement functionality and traceability in the Web3 era takes center stage. Examining how these elements contribute to the efficiency and security of cross-border transactions lays the groundwork for a more transparent and accountable financial ecosystem. Chapter 49 - Tracing Transactions: The Vital Role of Traceability: In the pursuit of a more accountable financial landscape, traceability emerges as a crucial theme. This chapter explores how traceability, enabled by cryptocurrency, becomes a cornerstone for fostering trust in cross-border transactions.

Chapter 50 - Balancing Confidentiality and KYC in Web3: As our exploration concludes, we delve into the balance between confidentiality and Know Your Customer (KYC) processes in Web3. Understanding this delicate equilibrium is essential for shaping a future where privacy and compliance coexist.

Part V stands as a testament to the   interplay between decentralized technologies and the global financial stage. Each chapter unfolds a layer of complexity, offering insights into the challenges, innovations, and promises that define the frontier of cross-border payments within the Web3 wilderness. Together, these chapters invite readers to contemplate the transformative potential that Web3 holds for the future of international transactions, setting the stage for the uncharted territories that lie ahead.

Chapter 41: Introduction to Cross Border Payments

In the dance of global commerce, the concept of cross-border payments takes center stage, orchestrating the movement of funds across national boundaries. This chapter embarks on an exploration of the fundamental s of cross-border payments, unraveling their essence, processes, and significance in the interconnected world of finance.

> *Cross Border Payments: Payments between any two parties in different jurisdictions, where the laws of both jurisdictions come into effect.*

Cross-border payments, at their core, constitute the financial transactions that traverse national borders.[208] They form the lifeblood of international trade, enabling businesses and individuals to engage in transactions beyond the confines of their domestic markets. In essence, these payments facilitate the movement of money between entities situated in different countries, underpinning the interconnectedness of the global economy.

Understanding the Basics: How Cross-Border Payments Work

At its most basic level, the mechanism of cross-border payments involves a series of steps. A party initiating an international transaction sends a payment instruction to their financial institution, which serves as the starting point of the journey. The financial institution then engages with corresponding institutions in the recipient's country to facilitate the transfer.

---

[208] Staff (2023). "What are Cross-Border Payments." ACI Worldwide at https://www.aciworldwide.com/cross-border-payment-processing.

Swift's global network plays a pivotal role in this process, providing a secure messaging system that ensures the swift and accurate transmission of payment instructions between financial institutions worldwide. However, traditional cross-border payment systems, while effective, often encounter challenges such as delays, high fees, and limited transparency.

In recent times, emerging technologies, including blockchain and decentralized finance (DeFi), have entered the arena, promising to revolutionize the landscape of cross-border payments.[209] These technologies aim to address the limitations of traditional systems by offering enhanced speed, reduced costs, and increased transparency in international transactions.

The Importance of Cross-Border Payments: Catalyzing Global Trade

The significance of cross-border payments extends beyond the mechanics of financial transactions. They are the lifeblood of international trade, fostering economic growth and globalization. Businesses leverage these transactions to engage in commerce across borders, tapping into new markets and forming global supply chains.[210]

Moreover, cross-border payments facilitate remittances, allowing individuals to support their families in different countries. This financial lifeline plays a crucial role in improving the economic well-being of communities around the world. The efficient functioning of cross-border payment systems is, therefore, essential for the sustenance and growth of economies on a global scale.

---

[209] Staff (2023). "Your Secure Guide to Cross-Border Payments." IR at https://www.ir.com/guides/cross-border-payments.

[210] IMF at https://www.elibrary.imf.org/view/journals/001/2023/072/article-A001-en.xml.

Conclusion: Setting the Stage for Deeper Exploration

In conclusion, this chapter serves as the opening act in our exploration of cross-border payments. As we traverse through the following chapters, we will delve into the intricacies, challenges, and innovations that shape the landscape of international transactions. The s of cross-border payments are not just a financial mechanism; they are the arteries that nourish the global economy. Our journey continues, promising a deeper understanding of this vital component in the world of finance.

Chapter 42: The Roll of Swift in Global Transactions

Introduction

In the realm of international finance, the story of cross-border payments is woven into the historical ecosystem of the Society for Worldwide Interbank Financial Telecommunication, better known as SWIFT. This chapter embarks on a journey through time, exploring the inception, evolution, and fundamental workings of SWIFT as a bulwark in global financial transactions.

SWIFT 1.0

The roots of SWIFT trace back to the early 1970s, a period marked by financial globalization and an increasing need for efficient communication among banks across borders. Prior to SWIFT, the absence of a standardized messaging system led to a fragmented and often cumbersome process for exchanging financial information.[211] Established in 1973, SWIFT emerged as a cooperative venture among banks to address these challenges. Its primary mission was to create a standardized and secure messaging network that would facilitate the rapid and reliable exchange of financial information globally. SWIFT aimed to streamline communication, reduce errors, and enhance the overall efficiency of cross-border transactions.[212]

At its core, SWIFT operates as a messaging network that enables financial institutions worldwide to communicate and exchange information securely.[213] The system employs a standardized set

---

[211] Staff (2023). "A Brief History of SWIFT." Information Builders at https://iwayinfocenter.informationbuilders.com/TLs/TL_soa_ebiz_swift/source/intro_swift9.htm.

[212] Staff (2023). "SWIFT History" SWIFT at https://www.swift.com/about-us/history.

[213] Lazo, Bernardo (2023). "50 Years of SWIFT" ADP at https://rethinkq.adp.com/history-of-swift/.

of message formats, known as SWIFT messages, which encapsulate various details related to financial transactions, including payment instructions, account details, and authentication data.

The process begins when a financial institution sends a SWIFT message to initiate a cross-border payment. The message navigates through the SWIFT network, reaching the recipient's financial institution, which then carries out the necessary steps to fulfill the payment.[214] SWIFT acts as a facilitator, ensuring that the communication is secure, standardized, and swiftly executed.

Over the years, SWIFT has evolved to meet the changing demands of the financial landscape. It has expanded its services beyond messaging to encompass various aspects of financial communication. SWIFT's standardized codes, such as Business Identifier Codes (BICs) and International Bank Account Numbers (IBANs), have become integral components in identifying banks and accounts, further streamlining cross-border transactions.[215]

While SWIFT has played a pivotal role in shaping the efficiency of cross-border payments, it has not been without challenges and criticisms. The system has faced scrutiny for issues such as delays, high costs, and concerns related to security.[216] The growing complexities of international finance have prompted discussions around the need for further innovation and improvement in cross-border payment systems.

---

[214] SWIFT (2023), *supra*.
[215] Scott, Susan & Zachariadis, Markos (2012), "Origins and Development of SWIFT." LSE at https://eprints.lse.ac.uk/46490/1/Origins%20and%20Development%20of.pdf.
[216] Staff (2023). "What is SWIFT? What to know about the international banking system." Stripe at https://stripe.com/resources/more/what-is-swift.

Conclusion

In conclusion, SWIFT stands as a stalwart in the realm of cross-border transactions, having weathered the tides of financial evolution for nearly five decades. Its standardized messaging system has been a bulwark in fostering communication among financial institutions worldwide. As we delve deeper into the intricacies of international finance, understanding the historical narrative of SWIFT sets the stage for unraveling the complexities that lie ahead.

Chapter 43: SWIFT 2.0- Innovations and Improvements

Introduction: The Dawn of SWIFT 2.0

As the winds of change swept through the landscape of global finance, SWIFT, the bulwark of cross-border transactions, embarked on a transformative journey with the advent of SWIFT 2.0. This chapter unravels the narrative of SWIFT's evolution, delving into the enhancements brought forth by SWIFT 2.0 and the impact on the ecosystem of international financial transactions.[217]

SWIFT 1.0, with its standardized messaging system, had long been the backbone of global financial communication. However, as the complexities of international finance deepened and the demand for faster, more transparent, and cost-effective transactions grew, SWIFT recognized the imperative for innovation. SWIFT 2.0 emerged as a response to these evolving s, aiming to address the limitations of its predecessor.[218]

Key Enhancements of SWIFT 2.0: Weaving a Finer Fabric of Connectivity

1. *Speed and Efficiency:* One of the hallmark improvements ushered in by SWIFT 2.0 was a significant boost in speed and efficiency. The upgraded platform streamlined processes, reducing the time required for cross-border payments. Transactions that once took days to settle could now be executed in near real-time, fostering a quicker and more responsive global financial ecosystem.

---

[217] Swift (2023), *supra*.

[218] Smart, Evander (2016). "Swift 2.0? Ripple Signs Major Banks to Blockchain Payment Deal." Coin Telegraph at https://cointelegraph.com/news/swift-20-ripple-signs-major-banks-to-blockchain-payments-deal.

2. *Enhanced Security Measures:* Recognizing the paramount importance of security in the digital age, SWIFT 2.0 implemented robust security measures. Advanced encryption protocols, multi-layered authentication, and continuous monitoring became integral components of the upgraded system, fortifying the resilience of cross-border transactions against potential threats.
3. *Integration of Emerging Technologies:* SWIFT 2.0 embraced the potential of emerging technologies such as blockchain and distributed ledger technology (DLT). The integration of these technologies into the SWIFT framework brought about increased transparency, traceability, and reduced friction in cross-border payments.
4. *Expanded Message Standards:* Building on the foundation of standardized messaging, SWIFT 2.0 introduced expanded message standards to accommodate the evolving needs of financial institutions. These standards facilitated richer data exchange, allowing for more nuanced and comprehensive information to be transmitted within the SWIFT network.

The enhancements introduced by SWIFT 2.0 had a profound impact on the global financial landscape. Financial institutions, businesses, and individuals engaged in cross-border transactions found themselves operating within a more responsive and technologically advanced ecosystem.[219] The newfound speed, security, and adaptability of SWIFT 2.0 positioned it as a cornerstone for the ever-expanding web of international finance.

---

[219] ADP, *supra*.

Conclusion: SWIFT 2.0 - A Chapter in Financial Evolution

In conclusion, SWIFT 2.0 emerged as an important stage in the ongoing evolution of global finance. The enhancements brought forth by this iteration addressed the challenges posed by an increasingly interconnected and financial world. As we traverse the intricacies of cross-border transactions, understanding the advancements of SWIFT 2.0 illuminates the path toward a more efficient, secure, and interconnected future in international finance.

Chapter 44: Ripple/XRP & SWIFT GPI- Attempting to Revolutionize Cross Border Transactions

Introduction: The Crossfire of Cross-Border Transactions

In the ever-evolving realm of cross-border transactions, two prominent players, SWIFT GPI and Ripple XRP, have garnered attention as potential architects of a new era in international finance. This chapter delves into the intricacies of SWIFT GPI and Ripple XRP, exploring their features, strengths, and the areas where they grapple with the imperatives of speed, cost, and efficiency in the global payment landscape.

SWIFT GPI: Transforming the Traditional ecosystem

SWIFT Global Payments Innovation, or SWIFT GPI, represents a significant stride in enhancing the conventional SWIFT system.[220] Launched to address the shortcomings of its predecessor, SWIFT GPI introduces features such as real-time tracking, end-to-end payment transparency, and quicker settlements, injecting a dose of modernity into the traditional ecosystem of cross-border transactions.

SWIFT GPI excels in providing improved visibility and traceability throughout the payment lifecycle. With its focus on transparency and speed, it has become a go-to solution for financial institutions seeking to streamline cross-border payments.[221] The ability to track payments in real time and offer a more predictable payment experience has been a hallmark of SWIFT GPI.

---

[220] SWIFT, *supra*.
[221] Staff (2023). "What is SWIFT GPI?" Currency wave at https://currencywave.com/what-is-swift-gpi/.

However, SWIFT GPI still grapples with certain challenges. The legacy infrastructure, though upgraded, retains some inherent complexities that can hinder the speed and efficiency of transactions.[222] Additionally, the cost-effectiveness of SWIFT GPI transactions may vary, and certain regions may experience limitations in the real-time tracking functionality.

Ripple XRP: Riding the Waves of Innovation

On the other side of the spectrum, Ripple XRP stands as a blockchain-based solution designed to revolutionize cross-border payments.[223] By leveraging distributed ledger technology, Ripple XRP aims to overcome the traditional constraints of international transactions, introducing speed, cost-effectiveness, and efficiency into the financial equation.

Ripple XRP shines in terms of transaction speed and cost efficiency. The decentralized nature of blockchain enables near-instant settlements and reduces intermediary costs, presenting a compelling alternative to traditional cross-border payment systems.[224] Ripple's commitment to fostering partnerships with financial institutions globally has propelled XRP into the spotlight as a disruptive force in the industry.

Despite its merits, Ripple XRP faces challenges. Regulatory uncertainties and the need for widespread adoption pose hurdles to its seamless integration into the global financial infrastructure. Moreover, the reliance on traditional banking

---

[222] SWIFT (2023B). "SWIFT GPI." At https://www.swift.com/our-solutions/swift-gpi.

[223] Sterling, Alex (2023). "Ripple Cross-Border Payments: Transforming International Transactions." Doubloin at https://www.doubloin.com/learn/ripple-cross-border-payments.

[224] Ripple Labs at https://ripple.com/solutions/cross-border-payments/.

relationships may introduce frictions in achieving the intended levels of decentralization.[225]

Conclusion

As we navigate the crossroads of the internet, some of the traditional cross-border payment options act as crossroads demons, exacting a fierce toll on those who want to use them. In the analysis of SWIFT GPI and Ripple XRP, it becomes evident that both solutions bring unique strengths to the table. The choice between them hinges on specific priorities and requirements of financial institutions and businesses engaged in cross-border transactions. The evolving landscape suggests that synergies between traditional systems and blockchain-based innovations may pave the way for a more harmonious future in global payments.

In conclusion, the comparison between SWIFT GPI and Ripple XRP underscores the transformative s shaping cross-border transactions. While each solution addresses certain pain points, neither stands as a panacea. The future lies in the exploration of symbiotic relationships, where the strengths of traditional systems and the innovations of blockchain converge to create a seamless, efficient, and cost-effective global payment ecosystem.

---

[225] Staff (2023). "RippleNet: A beginners guide to the decentralized network of banks." Coin Telegraph at https://cointelegraph.com/learn/ripplenet-the-decentralized-network-of-banks.

Chapter 45: General Crypto Utility in Cross Border Payments

Introduction: The Rise of Cryptocurrency in Global Transactions

In the evolving landscape of international payments, cryptocurrency has emerged as a disruptive force, challenging traditional systems like SWIFT. One of the key attributes that positions cryptocurrency as a game-changer is its self-validating nature. This chapter delves into the intricacies of how cryptocurrency's inherent validation mechanisms contribute to its efficacy in international payments, eclipsing the inefficiencies of outdated systems.

Cryptocurrency's Self-Validating Mechanism: A Paradigm Shift

Self-Validation Mechanisms: bits of code that allow cryptocurrencies to operate in a trustless system, validating each transaction that queries the smart contract.

Unlike traditional payment systems that rely on centralized authorities for validation, cryptocurrencies operate on decentralized networks. The decentralized nature ensures that transactions are verified through a consensus mechanism, often involving nodes or miners spread across the network.[226] This self-validating feature not only enhances security but also eliminates the need for a single point of control.

---

[226] Frankenfield, Jake (2023). "What are consensus mechanisms in Blockchain and Cryptocurrency." Investopedia at https://www.investopedia.com/terms/c/consensus-mechanism-cryptocurrency.asp.

Cryptocurrencies leverage blockchain technology, a distributed ledger that records transactions in a sequential and immutable manner. Once a transaction is validated and added to the blockchain, it becomes resistant to tampering or alteration.[227] This immutability ensures the integrity of transactions, providing a level of trust that is often elusive in centralized systems.

Smart contracts, self-executing agreements with the terms directly written into code, contribute to the self-validating nature of certain cryptocurrencies. These contracts automatically execute predefined actions when conditions are met. In the context of international payments, smart contracts enable automated validation, streamlining the transaction process and minimizing the need for intermediaries.[228]

Cryptocurrency vs. SWIFT: A Comparative Analysis

Traditional international payment systems, such as SWIFT, are characterized by multi-day settlement times and a series of intermediaries. Cryptocurrency transactions, in contrast, operate on a 24/7 basis and can achieve near-instant settlement.[229] The elimination of intermediaries not only accelerates the process but also reduces associated costs.

Cryptocurrencies provide transparent and traceable transactions due to the public nature of blockchain ledgers.[230] Each

---

[227] Staff (2023). "What is Immutable Ledger in Blockchain and its Benefits." Solu Labs at https://www.solulab.com/what-is-immutable-ledger-in-blockchain-and-its-benefits/.

[228] Linked In (2023). "How can Smart Contracts Streamline Cross-border payments." At https://www.linkedin.com/advice/0/how-can-smart-contracts-streamline-cross-border-payments-k1cwe.

[229] Basulto, Dominic (2023). "Better Buy: XRP or Solana?" Motley Fool at https://www.fool.com/investing/2023/11/30/better-buy-xrp-or-solana/.

[230] Adams, Josh (2024). "Transparency in Blockchain is a double edged sword, says DOP Head of Protocol." CCN at

transaction is recorded on the blockchain, and participants can track the entire history of a cryptocurrency unit. This transparency contrasts sharply with the opacity often associated with traditional banking systems.

Cryptocurrency transactions can be more cost-effective than traditional international transfers, especially for cross-border transactions involving different currencies.[231] Cryptocurrencies eliminate or minimize fees associated with currency conversion and intermediary banks, resulting in a more streamlined and cost-efficient process.

Conclusion

In conclusion, the self-validating nature of cryptocurrency stands as a testament to its efficacy in international payments. By leveraging decentralization, consensus mechanisms, and innovative features like smart contracts, cryptocurrency transcends the limitations of outdated systems like SWIFT. The inherent attributes of speed, transparency, and cost-effectiveness position cryptocurrency as a formidable contender in reshaping the landscape of global transactions, marking a paradigm shift toward a more efficient and inclusive financial future.

---

https://www.ccn.com/news/transparency-blockchain-double-edged-sword-dop/.

[231] Basulto, *supra*.

Chapter 46: Jurisdictional Challenges to Cross-Border Payments

Introduction: Navigating the Legal Maze of Cross-Border Transactions

Cross-border payments, despite their technological advancements, are entangled in a complex web of jurisdictional challenges. This chapter delves into the legal intricacies that often complicate international transactions, exploring how varying jurisdictions impact cross-border payments. Understanding these challenges is crucial for businesses and individuals seeking to navigate the global financial landscape.

The Complex Landscape of Jurisdictional Oversight

- *Divergent Regulatory Frameworks:* Different countries have distinct regulatory frameworks governing financial transactions, and this diversity poses a significant challenge for cross-border payments.[232] The absence of standardized regulations can lead to discrepancies in compliance requirements, making it difficult for parties involved to adhere to a unified set of rules.
- *Ambiguity in Cryptocurrency Regulation:* The rise of cryptocurrency has added a layer of complexity to cross-border transactions. Jurisdictions vary widely in their approach to regulating digital assets.[233] Some embrace cryptocurrencies, while others impose stringent restrictions. Navigating this diverse landscape requires a

---

[232] Staff (2023). "Regulatory Divergence: Costs, Risks, Impacts." IFAC at https://www.ifac.org/_flysystem/azure-private/publications/files/IFAC-OECD-Regulatory-Divergence.pdf.

[233] Staff (2023). "Cryptocurrency: Select Policy issues." Congressional Research Service at https://crsreports.congress.gov/product/pdf/R/R47425.

nuanced understanding of how each jurisdiction views and regulates cryptocurrencies.

- *Anti-Money Laundering (AML) and Know Your Customer (KYC) Compliance:* AML and KYC regulations vary globally, impacting cross-border payments that involve traditional and digital currencies. Ensuring compliance with these regulations becomes challenging when jurisdictions have disparate requirements. The lack of standardized AML and KYC practices can hinder the seamless flow of funds across borders.[234]

Challenges in Resolving Disputes

- *Legal Recourse and Enforcement:* In the event of disputes in cross-border transactions, legal recourse can become convoluted. Enforcing contracts and resolving conflicts may involve navigating the legal systems of multiple jurisdictions. The lack of a unified legal framework for cross-border payments exacerbates the challenges associated with dispute resolution.
- *Cross-Border Taxation Issues:* Taxation is a complex facet of cross-border transactions. Varying tax regulations in different jurisdictions can result in ambiguity and potential tax liabilities for businesses and individuals. Understanding and adhering to diverse tax laws is crucial to avoiding legal complications and financial penalties.

The Need for Harmonization and Global Cooperation

- *Standardization of Regulatory Practices:* To mitigate jurisdictional challenges, there is a growing need for the standardization of regulatory practices related to cross-

---

[234] Sav, Daria (2023). "AML/KYC Guide for Fintech Companies 2024." Sumscriber at https://sumsub.com/blog/fintech/.

border payments.[235] Collaborative efforts among countries and regulatory bodies can contribute to the development of a unified set of rules, fostering a more transparent and predictable environment.

- *International Cooperation on Cryptocurrency Regulation:* Given the global nature of cryptocurrencies, international cooperation is essential for creating a cohesive regulatory framework.[236] Establishing common ground on issues such as taxation, classification, and legal recognition of digital assets can streamline cross-border transactions involving cryptocurrencies.

Conclusion: Charting a Course Through Jurisdictional Complexities

In conclusion, the jurisdictional challenges in cross-border payments underscore the need for a concerted effort to harmonize regulatory practices globally. Businesses, financial institutions, and regulatory bodies must collaborate to create a more cohesive and standardized framework. Navigating the legal maze of cross-border transactions requires a nuanced understanding of jurisdictional differences, coupled with a commitment to fostering international cooperation. As the financial landscape continues to evolve, addressing these challenges becomes imperative for building a more inclusive and efficient global payment ecosystem.

---

[235] Insart at https://www.insart.com/signals/compliance-and-regulations-for-fintech.

[236] Langarde, Christine (2019). "The Next Steps for International Collaboration in Fintech." IMF at https://www.imf.org/en/News/Articles/2019/06/07/sp060819-lagarde-the-next-steps-for-international-cooperation-in-fintech.

Chapter 47: The Need for a Unified Regulatory Framework

Introduction

The challenges of cross-border payments are magnified by the lack of a unified regulatory system. In this chapter, we examine the compelling need for a harmonized approach to regulation, aiming to streamline and fortify the global landscape of cross-border transactions.

The Fragmented Regulatory Landscape

- *Regulatory Fragmentation's Impact:* The current state of cross-border payments is marked by a fragmented regulatory landscape.[237] Each jurisdiction operates with its own set of rules, creating a challenging environment for businesses and individuals engaged in international transactions. The absence of a unified framework hampers efficiency and breeds uncertainty.[238]
- *Compliance Challenges:* Businesses operating across borders grapple with diverse compliance requirements. Navigating through the regulatory intricacies of each jurisdiction demands substantial resources and a thorough understanding of local laws. A unified regulatory system would alleviate these challenges by providing a standardized set of rules applicable globally.

The Case for a Unified Regulatory System

- *Streamlining Compliance:* One of the primary advantages of a unified regulatory system is the

---

[237] Staff (2023). "Cryptocurrency Regulation: Striking a Balance Between Innovation and Compliance." Bankor at https://www.linkedin.com/pulse/cryptocurrency-regulation-striking-balance-between-innovation/.

[238] Langarde, *supra*.

streamlining of compliance processes. Businesses engaging in cross-border transactions would benefit from standardized rules, reducing the complexities associated with adhering to disparate regulatory frameworks.[239]

- *Enhancing Financial Inclusion:* A unified regulatory system could contribute to fostering financial inclusion globally.[240] Simplified and standardized regulations would encourage the participation of individuals and businesses from diverse regions, creating a more inclusive financial ecosystem.
- *Facilitating Innovation:* A harmonized regulatory environment would pave the way for greater innovation in the financial sector. Startups and established institutions alike could operate with more certainty, fostering the development and adoption of cutting-edge technologies and solutions in the cross-border payments space.[241]

Challenges and Considerations

- *Balancing National Interests:* Implementing a unified regulatory system requires a delicate balance between global cooperation and respect for national sovereignty.[242] Striking this balance is essential to

---

[239] Staff (2023). "How to Navigate the Evolving Landscape of Cryptocurrency Compliance." Complysci at https://www.complysci.com/resources/blog/how-to-navigate-the-evolving-landscape-of-cryptocurrency-compliance.

[240] *Id.*

[241] *Id.*

[242] Benigno, Pierpaolo, Schilling, Linda & Uhlig, Harald (2022). "Crytpocurrencies, Currency Competition and the Impossible Trinity." Journal of International Economics. At https://www.sciencedirect.com/science/article/pii/S0022199622000332.

ensure that the regulatory framework addresses global concerns while accommodating the unique needs and interests of individual nations.

- *Coordination among Regulatory Bodies:* Achieving a unified regulatory system necessitates effective coordination among regulatory bodies worldwide. Collaborative efforts, information sharing, and mutual understanding are critical components of building a cohesive framework that transcends borders.

Conclusion: Building a Future of Seamless Transactions

The need for a unified regulatory system for cross-border payments is increasingly evident. Such a system would not only simplify compliance processes but also foster financial inclusion and innovation. While challenges exist, the benefits of a harmonized approach far outweigh the complexities. The call for unity in regulation resonates as a pivotal step toward building a future where cross-border transactions are seamless, transparent, and accessible to all.

Chapter 48: Settlement Functionality and Security

Introduction

Settlement functionality stands as a cornerstone in the world of cryptocurrency, cross-border payments, and the evolving landscape of Web3. In this chapter, we delve into the multifaceted aspects of settlement, exploring its critical role and significance in these domains.

> *Decentralization, Immutability, and the Ledger are elements which make for a strong cryptocurrency.*

Settlement in cryptocurrency refers to the finalization of transactions on a blockchain. Unlike traditional finance, where intermediaries oversee settlements, cryptocurrencies operate on decentralized networks, ensuring trustless and transparent confirmation of transactions.

The immutability of blockchain technology ensures that once a transaction is settled, it cannot be altered. Finality, a key characteristic, signifies that the settlement is irreversible, providing a high level of security and eliminating the need for intermediaries.

The Nexus of Settlement and Cross-Border Payments

- *Speed and Efficiency:* Settlement functionality plays a pivotal role in enhancing the speed and efficiency of cross-border payments.[243] Traditional settlement systems often involve delays due to intermediary banks

---

[243] Statista (2024) "Average transactions speeds of 74 cryptocurrencies with the highest market cap as of January 2024." at https://www.statista.com/statistics/944355/cryptocurrency-transaction-speed/.

and various clearing processes. Cryptocurrency, with its swift and direct settlement, addresses these inefficiencies.

- *Cost Reduction:* The direct settlement facilitated by cryptocurrencies contributes to cost reduction in cross-border transactions. With fewer intermediaries involved, associated fees and processing costs are minimized, making cross-border payments more economical.

Settlement in the Web3 Ecosystem

- *Smart Contracts and Automated Settlement:* Web3, characterized by decentralized applications (DApps) and smart contracts, leverages settlement functionality to automate and streamline various processes.[244] Smart contracts, self-executing pieces of code, enable programmable settlements based on predefined conditions, introducing a new paradigm of efficiency.
- *Interconnectedness of Systems:* Settlement in Web3 is not confined to financial transactions. It extends to the settlement of agreements, execution of code, and even governance processes within decentralized autonomous organizations (DAOs). The interconnected nature of Web3 systems relies heavily on efficient settlement mechanisms.

The Importance of Efficient Settlement

- *Trust and Reliability:* In both cryptocurrency and cross-border payments, efficient settlement mechanisms enhance trust and reliability.[245] Participants in

---

[244] Staff (2023). "Smart Contract Automation Explained." Chainlink at https://chain.link/education-hub/smart-contract-automation.

[245] Widjaya, Ivan (2023). "How Smart Contracts will change the way you do business." Noobpreneur at

transactions, whether financial or contractual, benefit from the assurance that settlements occur promptly and securely.

- *Catalyst for Innovation:* Efficient settlement processes act as a catalyst for innovation. The speed and cost-effectiveness introduced by cryptocurrency settlements have spurred further exploration of decentralized finance (DeFi) and novel use cases within the Web3 ecosystem.

Conclusion

In summary, settlement functionality serves as the *piece de resistance* in the realms of cryptocurrency, cross-border payments, and Web3. Its decentralized, swift, and cost-effective nature reshapes traditional paradigms, offering a glimpse into a future where transactions, agreements, and governance are settled seamlessly across interconnected systems. As settlement mechanisms continue to evolve, they lay the foundations for a digital landscape characterized by efficiency, trust, and boundless possibilities.

---

https://www.noobpreneur.com/2023/09/14/how-smart-contracts-will-change-the-way-you-do-business/.

Chapter 49: Traceability in Cross Border Transactions

Introduction

In the web of cross-border transactions, traceability emerges as a crucial aspect, ensuring transparency, accountability, and security. This chapter delves into the importance of traceability and how cryptocurrency, with its unique features, contributes to enhancing traceability in cross-border transactions.

Understanding Traceability in Cross-Border Transactions

- *Visibility Across Borders:* Traceability involves the ability to track and monitor transactions from initiation to completion.[246] In cross-border transactions, where funds traverse diverse jurisdictions, having a clear and visible trail becomes imperative for regulatory compliance and security.
- *Challenges in Traditional Systems:* Traditional cross-border payment systems often face challenges in traceability. Intermediaries, multiple clearing processes, and disparate systems can create opacity in tracking the movement of funds, leading to inefficiencies and potential vulnerabilities.[247]

Cryptocurrency's Contribution to Enhanced Traceability

- *Immutable Ledger:* Cryptocurrencies operate on blockchain technology, leveraging an immutable ledger where transactions are recorded in a tamper-resistant manner. This inherent feature significantly enhances

---

[246] Rosic, Ameer (2023). "Are Bitcoin Transactions Traceable? You will be surprised." BlockGeeks at https://blockgeeks.com/are-bitcoin-transactions-traceable/.
[247] *Id.*

traceability by providing a transparent and unalterable record of transaction history.[248]

- *Public Ledger and Address Transparency:* Many cryptocurrencies, including Bitcoin and Ethereum, utilize public ledgers that allow anyone to inspect transaction histories. Wallet addresses associated with transactions are pseudonymous but transparent, enhancing traceability while preserving user privacy.[249]
- *Smart Contracts for Automated Tracking:* Smart contracts, integral to various cryptocurrencies, enable the automation of certain processes, including tracking and triggering predefined actions based on specific conditions. This automation not only streamlines processes but also adds an additional layer of traceability.[250]

Benefits of Enhanced Traceability in Cross-Border Transactions

- *Regulatory Compliance:* Traceability is instrumental in meeting regulatory requirements. Cryptocurrency's transparent and auditable nature aids businesses and financial institutions in adhering to compliance standards, reducing the risk of illicit activities such as money laundering.
- *Security and Fraud Prevention:* A transparent and traceable system acts as a deterrent to fraudulent activities. The immutability of blockchain ensures that once a transaction is recorded, it cannot be altered, providing a secure and fraud-resistant environment.

---

[248] Haqshanas, Ruholamin (2023). "Immutable Ledger." Technopedia at https://www.techopedia.com/definition/immutable-ledger.
[249] *Id.*
[250] Widjaya, *supra*.

Cryptocurrency and Cross-Border Traceability in Practice

Cryptocurrency transactions, being digital and often decentralized, allow for real-time monitoring. This real-time aspect is particularly beneficial in cross-border scenarios where delays and discrepancies can have significant consequences.[251] Cryptocurrencies facilitate more efficient and traceable cross-border remittances. Individuals sending funds can trace the status of their transactions, ensuring that the intended recipients receive the funds promptly.

Conclusion

In conclusion, the importance of traceability in cross-border transactions cannot be overstated. Cryptocurrency, with its decentralized and transparent nature, emerges as a formidable solution to enhance traceability, addressing the challenges posed by traditional systems. As the digital landscape evolves, the marriage of cryptocurrency and traceability paves the way for a future where cross-border transactions are not just efficient but also conducted with a heightened level of transparency and accountability.

---

[251] *Id.*

Chapter 50: Confidentiality and KYC in Cross-Border Payments

Introduction

In the landscape of Web3, the balance between confidentiality and Know Your Customer (KYC) processes emerges as a critical consideration. This chapter explores the challenges and opportunities presented by this delicate equilibrium, highlighting the evolving paradigms of confidentiality and regulatory compliance in the decentralized realm.

Confidentiality in Web3: Upholding Privacy in a Decentralized World

1. **Decentralization Principles:** Web3, built on principles of decentralization, prioritizes user privacy. Confidentiality in transactions is a foundational element, empowering users with a level of anonymity not easily achievable in traditional systems.[252]
2. **Privacy Coins:** Cryptocurrencies designed with privacy features, often termed "privacy coins" like Monero and Zcash, go a step further in providing enhanced confidentiality. These coins utilize advanced cryptographic techniques to obfuscate transaction details.[253]

KYC in Web3: Navigating Regulatory Realities

1. **Regulatory Expectations:** Despite the decentralized ethos, Web3 entities are increasingly subject to

---

[252] Staff (2023). "The Future Privacy in a Web3 World. " Shardeum at https://shardeum.org/blog/privacy-in-web3/.
[253] *Id.*

regulatory frameworks.[254] KYC processes are vital in ensuring compliance with anti-money laundering (AML) and counter-terrorism financing (CTF) regulations.

2. **Identity Verification:** KYC processes in Web3 involve identity verification, aligning with traditional financial systems' standards.[255] Users may need to undergo verification procedures when interacting with decentralized finance (DeFi) platforms or participating in token sales.

Striking a Balance: Challenges and Opportunities

1. **Challenges of Absolute Anonymity:** While confidentiality is a key attraction, absolute anonymity can pose challenges, especially concerning illicit activities. Striking the right balance involves finding mechanisms that preserve user privacy without compromising regulatory expectations.
2. **Privacy-Enhancing Technologies (PETs):** The emergence of privacy-enhancing technologies offers solutions for achieving this delicate balance. Zero-knowledge proofs and ring signatures are examples of PETs that enable transaction validation without revealing sensitive information.

The Role of Decentralized Identity (DID):

1. **User-Centric Identity:** Decentralized Identity solutions aim to give users control over their identity information. Users can selectively disclose information, aligning with the principles of user-centric identity models.[256]

---

[254] Staff (2023). "Web3: The game-changer in European KYC data." eClear at https://eclear.com/article/web3-the-game-changer-in-european-kyc-data/.

[255] Toggle at https://www.togggle.io/web3-kyc-solution.

[256] Clarke, Anthony (2023). "What are Decentralized Identifiers (DID) and how will they boost web3." Nasdaq at

2. **Interoperability Challenges:** The interoperability of different KYC solutions and decentralized platforms remains a challenge.[257] Collaborative efforts to establish standards for decentralized identity could pave the way for smoother KYC processes across Web3.

Conclusion: Paving the Way for a Confident and Compliant Web3 Future

In conclusion, the delicate relationship between confidentiality and KYC in Web3 reflects the evolving nature of decentralized ecosystems. Balancing user privacy with regulatory compliance requires innovative solutions and a commitment to establishing industry-wide standards. As Web3 continues to mature, finding equilibrium becomes not just a challenge but an essential pursuit for fostering a secure, confidential, and compliant decentralized future.

---

https://www.nasdaq.com/articles/what-are-decentralized-identifiers-did-and-how-will-they-boost-web3.

[257] *Id.*

Conclusion Part V Navigating the Cross-Border Frontier in Web3

As we traverse the multifaceted landscape of cross-border payments within the Web3 domain, it becomes evident that the integration of decentralized technologies is reshaping the way we envision and conduct international transactions. The chapters encapsulated in Part V illuminate the challenges, innovations, and transformative potential embedded in the intersection of Web3 and global financial interactions.

Evolution of Cross-Border Payments in Web3: A Recap

1. **Introduction to Cross-Border Payments in Web3:** The journey commenced with a foundational exploration of cross-border payments in the Web3 era, unraveling the intricacies of international transactions within decentralized frameworks.
2. **Insights into SWIFT, SWIFT 2.0, and Ripple/XRP:** Chapters delving into SWIFT's history, the evolution of SWIFT 2.0, and the Ripple/XRP ecosystem offered a comprehensive understanding of the traditional and emerging mechanisms for cross-border transactions.
3. **Striking the Balance with Confidentiality and KYC:** The delicate balance between confidentiality and Know Your Customer (KYC) processes emerged as a central theme, reflecting the nuanced considerations in preserving user privacy while adhering to regulatory expectations.
4. **Settlement Functionality, Tractability, and Regulatory Compliance:** Exploring the importance of settlement functionality, traceability, and regulatory compliance provided insights into the pillars supporting efficient and secure cross-border transactions in the Web3 landscape.
5. **Challenges of Jurisdiction and the Call for Unified Regulation:** Chapters addressing jurisdictional

challenges and the need for a unified regulatory system underscored the complexities that arise when decentralized technologies intersect with the traditional legal and regulatory frameworks.

Themes Unveiled and Future Horizons:

1. **Balancing Act of Confidentiality and KYC:** The interplay between user privacy and regulatory compliance emerged as a central theme. Striking the right balance becomes pivotal for the sustainable growth of decentralized finance and cross-border interactions.
2. **Technological Solutions and Privacy-Enhancing Technologies:** The exploration of privacy coins, Privacy-Enhancing Technologies (PETs), and Decentralized Identity (DID) solutions highlighted the industry's proactive approach in developing innovative solutions to navigate the challenges posed by evolving regulatory landscapes.
3. **The Role of Decentralization:** Throughout these chapters, the enduring role of decentralization as a guiding principle in Web3 remained evident. The promise of a trustless, permissionless environment continues to drive advancements in global financial interactions.

Looking Forward: The Uncharted Frontiers of Web3 Cross-Border Payments:

1. **Interconnectedness and Collaboration:** The chapters collectively reinforce the notion that the future of cross-border payments in Web3 lies in interconnectedness and collaborative efforts. Industry players, regulators, and innovators must collaborate to establish standards that foster both innovation and compliance.
2. **Adapting to Regulatory Evolution:** As the regulatory landscape continues to evolve, businesses and

technologies in the Web3 space must remain adaptable. Proactive engagement with regulatory bodies can pave the way for a more inclusive and sustainable global financial ecosystem.

In conclusion, Part V has been a journey through the ecosystem of cross-border payments in the Web3 wilderness. From the historical foundations to the contemporary challenges, each chapter unraveled a layer of complexity, bringing to light the transformative potential and evolving s of cross-border interactions in the decentralized era. As we stand at the cusp of uncharted frontiers, the chapters in this part collectively echo the sentiment that Web3 is not just a technological evolution but a paradigm shift in how we conceptualize and execute cross-border transactions. The journey continues into the unexplored territories of Web3, where innovation, collaboration, and regulatory adaptation will shape the future contours of global finance.[258]

[258] In this section, we strove to keep this as concise as possible. Many who are looking at the legal phenomenon in cryptocurrency-specifically in relation to cross border payments- need the data in bullet format for ready access.

Part VI: Discussing the Challenges of Web3

Introduction Part VI

As we step into the realm of Part VI, a new chapter unfolds in our exploration of Web3 – a terrain marked by decentralization, cryptographic innovations, and transformative potential. This segment of our journey delves into the intricacies of regulatory landscapes, security imperatives, talent s, legacy integration challenges, resistance to change, and the perpetual quest for scalability. Each chapter within this part serves as a compass, guiding us through the and multifaceted challenges that define the Web3 wilderness.

The opening chapters confront the complex web of regulatory realities entwining Web3 technologies. Regulatory compliance, a necessary ingredient for stability, security, and wider adoption, stands as a formidable challenge. Our exploration will delve into the intricacies of global regulatory frameworks, seeking to understand the balance between the need for order and the necessity of fostering innovation in the decentralized landscape.

Security, a cornerstone of trust in the digital realm, emerges as a central theme in this part. We will scrutinize the vulnerabilities and threats faced by Web3 platforms, emphasizing the need for robust security measures. Beyond identifying risks, our journey will explore proactive strategies to fortify these decentralized ecosystems, ensuring the integrity of transactions and the confidentiality of user data.

A deep dive into the talent pool awaits, as we navigate the challenges of acquiring and nurturing skilled professionals in the Web3 space. The evolving demands of blockchain and decentralized technologies necessitate a workforce equipped with specialized knowledge. Chapters in this section will explore the strategies employed by organizations to attract, train, and retain the talent essential for steering the Web3 ship.
Infrastructure, the backbone of technological progress, poses

both opportunities and challenges in the Web3 universe. Our exploration will encompass the intricacies of aligning national, regional, and corporate technological infrastructures with the demands of a decentralized future. Strategic investments, collaboration, and adaptive approaches will be key focal points as we navigate the tech-infrastructure conundrum.

The clash between legacy systems and the evolving landscape of Web3 technologies takes center stage. Chapters will dissect the complexities of integration, exploring the dilemmas faced by organizations striving to blend the old with the new. Artificial intelligence emerges as a potential ally, smoothing the transition and mitigating challenges associated with legacy integration. Challenges extend beyond technological landscapes, delving into the psychology of change. Media narratives, often fueled by fear and misinformation, cast shadows on the path to widespread Web3 adoption. Our exploration will address the impact of mainstream media narratives and the resistance to change within business communities, aiming to illuminate the way forward.

Navigating the turbulent waters of market s and business communities' hesitance to embrace change presents a unique set of challenges. Chapters will dissect the influences shaping market competition and delve into the psychological barriers hindering the acceptance of Web3 technologies within business circles. The odyssey concludes with an exploration of scalability, an ever-elusive goal in the realm of Web3. The challenge of balancing efficiency with security takes center stage as we delve into the technological advancements and adaptive strategies required to meet the increasing demands placed on decentralized systems.

As we embark on this Web3 odyssey through regulatory landscapes, security realms, talent ecosystems, technological infrastructures, legacy integration challenges, societal perceptions, markets, and scalability frontiers, each chapter within this part contributes to the collective understanding of the challenges and opportunities defining the decentralized future. The narratives unfold, inviting stakeholders, enthusiasts, and innovators alike to join in the dialogue and forge a path forward in the and ever-evolving world of Web3.

Chapter 51: Regulatory Compliance

Introduction

In the uncharted waters of Web3, where decentralization meets financial landscapes, the question of regulatory compliance emerges as a significant navigational challenge. The promises of blockchain, cryptocurrency, and decentralized systems are entangled with the complexities of existing regulatory frameworks, prompting a nuanced exploration of the hurdles and opportunities that regulatory compliance presents in the Web3 space.

A Regulatory Landscape to Explore

> *Web3: a vision for the internet where anything you can imagine can be tokenized, giving people access to their holdings anywhere they have internet access.*

As businesses and individuals traverse the Web3 seas, the absence of a unified regulatory framework poses challenges. Traditional regulatory structures designed for centralized systems struggle to encapsulate the decentralized, borderless nature of Web3 technologies.[259] This chapter delves into the intricacies of regulatory compliance, dissecting the various facets that contribute to the evolving landscape.

One of the primary issues in Web3 regulatory compliance lies in the decentralized and pseudonymous nature of transactions.[260]

---

259 Staff (2023). "The top legal challenges in Web3 Gaming you need to know." Bull Blockchain Law LLP at https://bullblockchainlaw.com/the-top-legal-challenges-in-web3-gaming-you-need-to-know/.

260 Li (Chong), Max (2023). "Absolute Privacy in Web3 is an Illusion." Forbes Digital Assets at https://www.forbes.com/sites/digital-

Traditional regulatory mechanisms, accustomed to clear ownership and identifiable entities, find themselves grappling with the anonymity that certain Web3 systems afford. This creates a conundrum where compliance officers face the challenge of ensuring adherence to laws without compromising the fundamental tenets of decentralization.

Smart contracts, a hallmark of Web3 functionality, add a layer of complexity to regulatory compliance. As self-executing contracts written into code, they operate without the need for intermediaries.[261] However, this autonomy raises questions about legal enforceability and liability. Navigating the regulatory ambiguity surrounding smart contracts becomes a crucial aspect of compliance in the Web3 ecosystem.

The global nature of Web3 technologies exacerbates the challenge of regulatory compliance. Varying regulatory approaches and attitudes toward decentralized technologies create a patchwork of legal landscapes. Companies operating in the Web3 space must navigate these disparities, understanding the implications of cross-border transactions on compliance requirements.

Conclusion

In the conclusion, it becomes evident that the regulatory seas of Web3 are both tumultuous and promising. The challenges posed by the clash between decentralized principles and traditional regulatory structures are real, but they also underscore the transformative potential of Web3 technologies. As the journey through the Web3 regulatory landscape unfolds,

---

assets/2023/10/25/absolute-privacy-in-web3-is-an-illusion/?sh=58f52ee24168.

[261] Staff (2023). "Smart Contracts and Proof-of-Stake: New Frontiers for Compliance." Inreg at https://www.innreg.com/blog/regulating-smart-contracts-and-proof-of-stake-new-frontiers-for-compliance.

businesses and regulators alike must collaborate to chart a course that fosters innovation while upholding legal and ethical standards. The next chapters will further illuminate the strategies and considerations essential for businesses navigating these regulatory waters, providing a compass for those venturing into the uncharted territories of Web3.

Chapter 52: Security Concerns

Introduction:

As businesses and individuals embark on the transformative journey into the Web3 era, the question of security looms large. The decentralized and interconnected nature of Web3 technologies introduces a unique set of security concerns that demand attention. This chapter delves into the multifaceted landscape of Web3 security, exploring the challenges and ongoing efforts to fortify digital ecosystems against evolving threats.

The Web3 ecosystem, built on principles of decentralization and trustlessness, faces a spectrum of security challenges.[262] One prominent concern is the susceptibility of decentralized networks to novel attack vectors. Smart contract vulnerabilities, consensus algorithm weaknesses, and the pseudo-anonymous nature of transactions contribute to a complex security landscape. In this chapter, we navigate through these challenges, shedding light on the vulnerabilities that businesses must be vigilant against.

Decentralization and Security:

While decentralization is a fundamental concept of Web3, it introduces a paradigm shift in security s. Traditional centralized systems often rely on a single point of control, allowing for easier implementation of security measures. In contrast, the distributed nature of Web3 makes it imperative to secure multiple nodes and prevent attacks across a network of participants.[263] This section explores how the principles of

---

[262] Marin, Samantha (2021). "What is a Trustless System?" BanklessDAO at https://medium.com/bankless-dao/what-is-a-trustless-system-3ded568c8921.

[263] Bitton, David (2023). How Web3 and Decentralization are Transforming the Real Estate Industry." Forbes Technology Council at https://www.forbes.com/sites/forbestechcouncil/2023/02/08/how-

decentralization impact the strategies employed to secure Web3 platforms.

Smart Contract Auditing and Best Practices:

Smart contracts, the self-executing code governing many Web3 transactions, are susceptible to exploitation if not adequately audited. Security breaches, often arising from vulnerabilities in smart contract code, can have far-reaching consequences. The chapter examines the importance of rigorous smart contract auditing and the evolving best practices adopted by developers to enhance the security of decentralized applications (DApps) and protocols.

Encryption and Privacy Concerns:

In the interconnected Web3 landscape, ensuring the privacy and security of user data becomes paramount. This section delves into the cryptographic protocols and encryption methods employed to safeguard sensitive information.[264] It explores the delicate balance between privacy and transparency, emphasizing the need for robust encryption measures to prevent unauthorized access to personal data.

Ongoing Security Initiatives:

The Web3 community is not passive in the face of these security challenges. Initiatives and collaborations are underway to fortify the foundations of Web3 technologies. From decentralized identity solutions to consensus algorithm upgrades, the chapter provides insights into the proactive measures being taken to

---

web3-and-decentralization-are-transforming-the-real-estate-industry/?sh=28030b6c6115.

[264] Mavrou, Iliana (2024). "How Web3 is Redefining Privacy and the Internet." Techopedia at https://www.techopedia.com/how-web3-is-redefining-privacy-on-the-internet.

address security concerns.[265] It highlights the importance of community-driven efforts in creating a resilient and secure Web3 ecosystem.

Conclusion

The journey through the Web3 security landscape reveals an evolving terrain. The challenges posed by decentralization, smart contract vulnerabilities, and privacy considerations necessitate a comprehensive approach to security. As businesses navigate this uncharted territory, understanding and actively participating in the ongoing initiatives to enhance Web3 security will be pivotal. The subsequent chapters will further explore specific strategies and emerging technologies that contribute to a more secure and resilient Web3 future.

---

[265] Weston, Georgia (2023). "Web3 Security: Protecting Your Digital Assets." 101 Blockchains at https://101blockchains.com/web3-security/.

Chapter 53: Talent Acquisition

Introduction:

The fast-paced evolution of Web3 technologies demands not only adaptation but also a deep pool of expertise. In this chapter, we explore the critical aspects of talent acquisition and training, both internal and external, as businesses strive to navigate and thrive in the Web3 marketplace.

Talent Acquisition in the Web3 Era:

As the Web3 ecosystem expands, the demand for skilled professionals with expertise in blockchain, decentralized technologies, and related domains intensifies.[266] The chapter delves into the nuances of talent acquisition, emphasizing the importance of recruiting individuals with a keen understanding of distributed ledger technologies, smart contracts, and decentralized applications. Businesses aiming to establish a foothold in the Web3 space must strategically identify and attract talent equipped to navigate its landscape.

Training Initiatives:

Equipping existing teams with the knowledge and skills necessary for Web3 adoption is a crucial element of successful integration. Internal training initiatives play a pivotal role in upskilling employees, fostering a culture of continuous learning, and ensuring that the workforce remains adept at leveraging decentralized technologies. The chapter explores various internal training models and the benefits they bring to organizations seeking to build Web3 expertise from within.

---

[266] Staff (2023). "The Wolr of Web 3 in Decentralized Talent Acquisition and Recruitment." DS Legends Pte. LTE at https://dslsingapore.medium.com/the-role-of-web3-in-decentralized-talent-acquisition-and-recruitment-82e697b3e79f.

In the realm of Web3, external training programs serve as invaluable resources for businesses and individuals alike. These programs, often offered by industry-leading organizations, educational institutions, or through collaboration with Web3 projects, provide comprehensive insights into blockchain, smart contracts, decentralized finance (DeFi), and other key components.[267] The chapter discusses the significance of external training in empowering professionals to grasp the intricacies of Web3 technologies and apply them effectively.

Web3 technologies are in a constant state of evolution, demanding a commitment to continuous learning. The chapter emphasizes the need for businesses to cultivate a culture that encourages ongoing education and adaptation. This includes providing access to up-to-date resources, supporting participation in industry events, and fostering communities of practice where knowledge sharing can thrive.

Strategic Partnerships and Collaboration:

Collaboration with external entities, including universities, research institutions, and industry consortia, can significantly enhance talent development initiatives. The chapter explores the benefits of strategic partnerships in creating pathways for talent acquisition and development. By collaborating with educational institutions and industry leaders, businesses can actively contribute to the cultivation of expertise in the broader Web3 community.[268]

---

[267] Inspired (2023). "Web 3.0 and Semantic Technologies- Introduction for Business. Inspired at https://dslsingapore.medium.com/the-role-of-web3-in-decentralized-talent-acquisition-and-recruitment-82e697b3e79f. https://www.inspired.org/training-web3-for-business.

[268] Stelzner, Michael (2023). "How to create Web3 Brand Partnerships." Social Media Examiner at https://www.socialmediaexaminer.com/how-to-create-web3-brand-partnerships//

Effectively measuring the success of talent acquisition and training programs is essential for optimizing strategies. The chapter provides insights into key performance indicators (KPIs) and metrics that businesses can use to evaluate the impact of their initiatives. From increased productivity to successful project implementations, understanding the return on investment (ROI) of talent development efforts is crucial for long-term success.

Conclusion

In conclusion, talent acquisition and training form the bedrock of success in the Web3 marketplace. Whether through strategic hiring, internal upskilling, or external training collaborations, businesses must prioritize the cultivation of expertise to navigate the complexities of decentralized technologies. As we move forward, the ability to attract, retain, and continuously develop a skilled workforce will be a defining factor in the success of organizations embracing the Web3 revolution.

Chapter 54: Closing the Gap: Advancing Technological Infrastructure in the Digital Era

Introduction:

In the ever-expanding digital universe, the challenge of aligning national, regional, or corporate technological infrastructure with the rapid pace of development is a formidable task. This chapter delves into the complexities of bridging this gap, exploring the hurdles, strategies, and the critical role of technological infrastructure in shaping the digital future.

The Pace of Digital Evolution:

The relentless speed at which the digital landscape evolves poses a significant challenge to existing technological infrastructures. From the rise of Web3 technologies to the integration of artificial intelligence and blockchain, the demands on infrastructure have never been greater. The chapter begins by examining the forces driving the digital evolution and the implications for outdated or insufficient technological frameworks.

> *Digital Evolution: The concept that the digital world evolves like the physical world and that the evolutions are based on stimuli.*

National and Regional Disparities:

Disparities in technological infrastructure between nations and regions compound the challenges of development. The chapter explores how varying levels of technological advancement create a digital divide, affecting not only economic competitiveness but also the ability of countries and regions to participate fully in the digital economy.[269] Case studies and examples illustrate the stark contrasts in infrastructure development and the subsequent impacts on societal progress.

Within the corporate landscape, the need for advanced technological infrastructure is equally pressing. The chapter discusses how businesses, regardless of size or industry, face the imperative of keeping pace with digital innovations to remain competitive. It explores the impact of outdated infrastructure on operational efficiency, customer experience, and the ability to adapt to changing markets.

Hurdles to Technological Advancement:

Identifying the hurdles to technological advancement is crucial for developing effective strategies. The chapter addresses common challenges, including outdated legacy systems, cybersecurity concerns, budget constraints, and the skills gap.[270] Each obstacle is examined in-depth, highlighting its implications for the broader goal of advancing technological infrastructure.

Amidst the challenges, the chapter explores strategies to catalyze progress in technological infrastructure. It emphasizes the importance of visionary leadership, public-private collaborations, and long-term investment in research and development. Case studies showcase successful examples where nations, regions, or corporations have effectively navigated the complexities of infrastructure transformation.

---

[269] Segel, Liz & Hatami, Homayoun (2023). "World Wide Web3." McKinsey & Company at https://www.mckinsey.com/~/media/mckinsey/email/onpoint/2023/11/2023-11-08c.html.

[270] Staff (2023). "Striking the balance in Web3 Social Interaction Across Diverse Cultures." IntoVerse at https://medium.com/@intoverse/striking-the-balance-in-web3-social-interaction-across-diverse-cultures-e3304e9b4322.

The Role of Emerging Technologies:

Web3 technologies, artificial intelligence, and blockchain emerge as catalysts for transforming technological landscapes. The chapter discusses how these innovations can be leveraged to leapfrog traditional stages of infrastructure development, providing efficient, scalable, and secure solutions. Real-world applications demonstrate the transformative power of integrating emerging technologies into existing infrastructures.[271]

Resilience and adaptability are identified as key characteristics of future-proof infrastructure. The chapter explores how building infrastructure capable of responding to unforeseen challenges, such as global pandemics or cybersecurity threats, is essential for long-term success. It advocates for an approach that combines flexibility, scalability, and a commitment to continuous improvement.

Conclusion:

In summary, the challenge of aligning national, regional, or corporate technological infrastructure with the digital universe is a multifaceted endeavor. The chapter underscores the urgency of addressing this challenge and provides insights into the transformative potential of embracing advanced technologies. As we navigate the complexities of the digital landscape, the ability to close the infrastructure gap will be pivotal in shaping a future where technological progress is inclusive and sustainable.

---

[271] Staff (2023). "Integrating Emerging Technologies into Engineering Practices in Emerging Economies." Utilities One at https://utilitiesone.com/integrating-emerging-technologies-into-engineering-practices-in-emerging-economies.

Chapter 55: Integration Costs

Introduction:

Decentralized Finance (DeFi) is heralded as a transformative force in the financial landscape, promising increased accessibility, efficiency, and transparency. However, this chapter delves into a critical aspect often overshadowed – the costs associated with integrating DeFi programs at the national, regional, and corporate levels. Understanding these costs is paramount for informed decision-making in the pursuit of a decentralized financial future.

The Multi-Level Perspective:

At the national level, embracing DeFi involves a strategic overhaul of financial frameworks. The chapter examines the costs governments may incur during the transition, including regulatory adaptations, the development of new compliance standards, and potential investments in blockchain infrastructure.[272] While DeFi offers the promise of financial inclusion and reduced reliance on traditional banking, the costs of reshaping national financial systems are complex and multifaceted.

Regional bodies and organizations play a pivotal role in navigating the integration of DeFi. This section explores the costs associated with coordinating regulatory efforts, fostering collaboration between neighboring nations, and addressing cross-border challenges. The chapter highlights potential expenses related to harmonizing legal frameworks and establishing regional DeFi infrastructure to ensure seamless cross-border transactions.

For businesses, the allure of DeFi lies in streamlined processes, reduced transaction costs, and improved financial inclusivity.

---

[272] Wealwin, *supra*.

However, the integration journey is not without its expenses. The chapter discusses the costs companies may face when adopting DeFi technologies, including initial investments in blockchain infrastructure, staff training programs, and potential adjustments to business models. Balancing the immediate costs with long-term benefits becomes a critical aspect of corporate decision-making.

The Technology, Regulatory & Training Investment:

A significant portion of the costs associated with DeFi integration lies in technology investments. The chapter explores the need for robust blockchain infrastructure, smart contract development, and secure decentralized applications (dApps).[273] The discussion emphasizes how technological investments are pivotal in unlocking the full potential of DeFi, ensuring scalability, security, and user-friendly interfaces.

Navigating the regulatory landscape is a substantial aspect of DeFi integration. The chapter examines the costs related to compliance with evolving regulations, legal consultations, and the potential need for specialized compliance teams. While DeFi strives for decentralization, the reality is that regulatory compliance remains a crucial aspect to gain legitimacy and trust on a global scale.

DeFi adoption introduces operational challenges that organizations must address. The chapter discusses the costs associated with restructuring internal processes, integrating DeFi protocols, and training personnel. The importance of upskilling the workforce to navigate decentralized technologies

---

[273] Powell, Jeff (2022). "Blockchain, Web3 and the potential for government transformation." Route Fifty at https://www.route-fifty.com/emerging-tech/2022/07/blockchain-web3-and-potential-government-transformation/375149/.

is emphasized, highlighting the investment required for a smooth transition.

Conclusion:

In conclusion, embracing DeFi is an adventure that involves costs at the national, regional, and corporate levels. The chapter underscores the need for a comprehensive understanding of these costs to make informed decisions. While the benefits of decentralized finance are substantial, navigating the complexities and financial commitments is essential. Striking a balance between immediate costs and the transformative potential of DeFi will be pivotal for nations, regions, and businesses looking to position themselves at the forefront of the decentralized financial revolution.

Chapter 56: Navigating the Legacy Labyrinth in Web3: AI as the Guiding Light

Introduction:

The integration of Web3 technologies into existing legacy systems presents a formidable challenge for businesses. In this chapter, we delve into the intricacies of this complex process, exploring the hurdles associated with legacy integration in the Web3 landscape. Furthermore, we examine how Artificial Intelligence (AI) emerges as a crucial ally in overcoming these challenges, paving the way for a seamless transition.

The Legacy Conundrum:

Legacy Systems: Computer systems optimized for Web 2.0 or the original internet still being used today, mostly by governments.

Legacy systems, characterized by their rigid structures and often outdated technologies, pose a significant obstacle to the swift adoption of Web3.[274] The chapter delves into the complexities businesses encounter when attempting to align these legacy frameworks with the decentralized,   nature of Web3. Issues such as incompatible data formats, disparate protocols, and security concerns take center stage as organizations grapple with a digital era that demands flexibility.

Challenges in Integration:

The challenges associated with integrating Web3 into legacy systems are multifaceted. We explore the resistance to change within traditional systems, the potential disruption to existing

[274] Barney, Nick (2023). "Legacy Systems." Tech Target at https://www.techtarget.com/searchitoperations/definition/legacy-application.

operations, and the financial investments required to update infrastructure. The chapter emphasizes the need for businesses to navigate these challenges strategically to unlock the transformative potential of Web3 without jeopardizing operational continuity.[275]

AI as the Integration Catalyst:

Artificial Intelligence emerges as a potent solution to streamline the integration of Web3 with legacy systems. The chapter discusses how AI-driven tools can analyze, interpret, and bridge the gaps between different data formats and protocols.[276] Through machine learning algorithms, AI can adapt to the intricacies of legacy structures, facilitating a smoother integration process. One of the critical aspects of legacy integration is the migration and transformation of vast amounts of data. The chapter explores how AI-powered algorithms can automate data migration, ensuring that valuable information from legacy systems seamlessly transitions into the decentralized Web3 environment.[277] The adaptability of AI is crucial in managing the diversity of data formats encountered during integration.

Security concerns often act as a deterrent in legacy system integration. The chapter discusses how AI can fortify cybersecurity measures by autonomously identifying vulnerabilities, predicting potential threats, and ally adapting security protocols. The self-learning capabilities of AI contribute

---

[275] Bhatt, Tuhin (2023). "Challenges of Legacy System Migration in Softeware Development + How to Overcome them." Intelivita at https://www.intelivita.com/blog/legacy-system-migration-challenges/.
[276] Sapra, Yatin (2023). "How AI is Making Web3 More accessible and User-Friendly." Hash Studios at https://hashstudioz.com/blog/how-ai-is-making-web3-more-accessible-and-user-friendly/.
[277] Staff (2023). "AI in Web3: How AI Manifests in the World of Web3." Leeway Hertz at https://www.leewayhertz.com/ai-in-web3/.

to creating a robust defense mechanism against evolving cyber threats.

The integration process impacts day-to-day operations, and here, AI plays a pivotal role in maintaining and enhancing overall efficiency. The chapter highlights how AI-driven automation can optimize processes, reducing manual intervention and ensuring a smooth transition. By identifying areas for process improvement, AI becomes a catalyst for operational excellence amid the transformative shift to Web3.[278]

Investing in AI for legacy integration prompts a discussion on cost considerations. The chapter explores how businesses can weigh the upfront costs of AI adoption against the long-term benefits, emphasizing the potential return on investment (ROI) through improved efficiency, reduced operational disruptions, and enhanced security.

Conclusion:

In conclusion, the integration of Web3 with legacy systems is a complex journey fraught with challenges, but AI emerges as a guiding light. The chapter underscores the transformative role of AI in mitigating integration complexities, optimizing operations, and fortifying cybersecurity. As businesses navigate the legacy labyrinth in the era of Web3, embracing AI-driven solutions becomes not just a strategic choice but a necessity for unlocking the full potential of decentralized technologies.

[278] *Id.*

Chapter 57: Navigating the Web3 Narrative: Overcoming Fear through Education

Introduction:

In the understanding of Web3, mainstream media plays a crucial role in shaping public perceptions. This chapter delves into the impact of media narratives on the Web3 community, exploring how fear and misinformation can impede progress. Furthermore, it advocates for active participation by companies in education programs to foster a better understanding of Web3, ultimately paving the way for increased consumer adoption.

Media Influence on Web3 Perception:

Mainstream media often amplifies sensational narratives that focus on the risks and uncertainties associated with Web3 technologies. The chapter discusses how fear-inducing stories can create apprehension among the general public, hindering the widespread acceptance and adoption of decentralized technologies. Misconceptions surrounding topics like cryptocurrency, blockchain, and decentralization can lead to resistance and reluctance.[279]

Fear-driven narratives pose challenges for the Web3 community, ranging from regulatory scrutiny to public skepticism. The chapter explores how these challenges can impede innovation, discourage investment, and limit the development of decentralized technologies. Overcoming these obstacles requires a concerted effort to address and rectify the misinformation circulating in the public sphere.

To counteract fear and misinformation, the chapter emphasizes the importance of education programs initiated by companies

---

[279] Staff (2023). "Understanding Web3 and its Security Implications." Cloud Flare at https://www.cloudflare.com/the-net/web3-security/.

involved in the Web3 space. These programs serve as a proactive approach to inform the public about the principles, benefits, and potential risks of Web3 technologies. By fostering a nuanced understanding, education becomes a powerful tool in dismantling unfounded fears.

Active Company Participation:

The system needs to advocate for active participation by Web3 companies in educational initiatives. Companies are encouraged to take a leading role in shaping the narrative around their technologies, providing accurate information, and demystifying complex concepts.[280] Engaging with the public through webinars, workshops, and informative content helps build trust and transparency.

Collaboration across the Web3 industry is crucial for achieving a collective impact in dispelling fear and fostering understanding. The chapter explores how joint efforts, involving multiple companies and stakeholders, can amplify the reach and effectiveness of educational programs. By presenting a united front, the Web3 community can counteract negative narratives with well-informed, positive messages.

Companies operating in the Web3 space are encouraged to embrace transparency and accountability as pillars of their communication strategies.[281] The chapter discusses how open and honest communication about the potential benefits and risks of Web3 contributes to building credibility. Transparency

---

[280] Kuntz, Gehrig (2023). "Web3 Skeptics and believers both need a reality check." Stack Overflow at https://stackoverflow.blog/2022/05/25/web3-skeptics-and-believers-both-need-a-reality-check/.

[281] Mitchell, Nick (2023). "OKX P2P Trading: A Comprehensive Guide to Success." Coin Clarity at https://coinclarity.com/okx-p2p-trading/ (Example).

establishes a foundation of trust, mitigating the impact of fear-driven narratives.

The Case of "CoinClarity"

In the world of cryptocurrency, knowledge is power. Navigating the volatile and jargon-filled landscape can leave even tech-savvy individuals feeling lost and overwhelmed. "CoinClarity," a revolutionary education project launched in 2020, has tackled this challenge head-on, empowering thousands of individuals to enter the crypto scene with confidence and make informed investment decisions.[282]

Demystifying the Digital Jungle:

CoinClarity's success stems from its comprehensive and accessible approach to education. Recognizing the diverse needs of learners, they offer:

- **Multi-format learning:** Engaging video tutorials, interactive quizzes, detailed blog posts, and live Q&A sessions cater to different learning styles and preferences.
- **Beginner-friendly approach:** Complex concepts are broken down into clear, jargon-free explanations, ensuring no one gets left behind.
- **Holistic curriculum:** The program covers the full spectrum of crypto knowledge, from blockchain fundamentals and market analysis to trading strategies and security practices.
- **Community focus:** A vibrant online forum and social media presence foster a supportive learning environment where individuals can connect, share experiences, and ask questions.

---

[282] https://bittrust.org/coin-clarity.

Beyond anecdotal evidence, CoinClarity's success is quantifiable:

- **Over 100,000 registered users** from across the globe, representing a diverse range of demographics and backgrounds.[283]
- **78% user satisfaction rate** based on feedback surveys and community engagement metrics.[284]
- **Significant increase in user confidence:** Pre-and post-program surveys reveal a dramatic shift in users' understanding and comfort level with cryptocurrency.[285]
- **Active alumni network:** Graduates remain engaged in the community, mentoring newer members and contributing to knowledge sharing.

Moving Beyond Education:

CoinClarity's ambition extends beyond education. They actively contribute to the broader crypto ecosystem through:

- **Collaborations:** Partnering with industry leaders to develop informative webinars, workshops, and educational resources.[286]
- **Advocacy:** Promoting responsible and ethical practices within the crypto community.
- **Innovation:** Exploring new learning technologies and developing interactive tools to enhance the educational experience.[287]

---

[283] Staff (2024). "Coin Clarity: Coin Clarity Helps you find your new favorite digital coins, exchanges, and wallets." Alternative to at https://alternativeto.net/software/coin-clarity/about/.

[284] *Id.*

[285] *Id.*

[286] Staff (2024). "Spot Bitcoin ETF 'A Step Towards Regulatory Clarity." Coin Desk at https://www.coindesk.com/tv/first-mover/spot-bitcoin-etfs-a-step-towards-regulatory-clarity-crypto-policy-outlook-2024/.

[287] *Id.*

The Future of Crypto Education:

CoinClarity's success sets a new standard for cryptocurrency education. It demonstrates that empowering individuals with knowledge and building a supportive community are key to fostering healthy and sustainable growth in the crypto space. As the industry evolves, we can expect CoinClarity to remain at the forefront, continuously adapting and innovating to meet the ever-changing needs of its learners.

The future of crypto education is bright, thanks to pioneers like CoinClarity. With their dedication to demystifying the digital jungle and empowering individuals with the tools they need to succeed, they pave the way for a more informed and confident crypto future for all.

Conclusion:

In conclusion, the Web3 community faces a significant challenge in mitigating the impact of fear-driven narratives perpetuated by mainstream media. By actively participating in education programs, companies can take a leading role in shaping a positive narrative, dispelling myths, and fostering understanding. Through transparency, collaboration, and accountability, the Web3 community can empower the public to embrace the transformative potential of decentralized technologies, ultimately paving the way for widespread adoption.

Chapter 58: Market Competition

Introduction:

As Web3 technologies strive for widespread adoption, markets play a pivotal role in shaping the landscape. This chapter delves into the impact of market competition on Web3 adoption, specifically addressing how legacy systems, keen on maintaining relevance, resort to dropping prices at a loss. The discussion explores the challenges posed by this competitive strategy and proposes strategies for navigating these complexities.

Market Competition and Legacy Systems:

The advent of Web3 technologies has introduced a new era of decentralized and innovative solutions. However, the chapter acknowledges that the transition faces resistance from established legacy systems. Faced with the threat of obsolescence, traditional systems resort to aggressive pricing strategies to retain market share, often leading to a race to the bottom.[288]

The chapter examines the phenomenon of price wars, where legacy systems strategically reduce prices to levels unsustainable for profitability. This tactic, known as using "loss leaders," allows traditional players to maintain competitiveness in the short term but can have detrimental long-term effects on the entire market, including Web3 adoption.

Impact on Web3 Startups and Innovators:

Web3 startups and innovators face unique challenges in a market environment dominated by legacy systems engaging in aggressive pricing. The chapter discusses how the pricing

---

[288] Hooper, Anatol (2023). "Law 3.0- an evolution toward on-chain legal resolution in web3." Coin Telegraph at https://cointelegraph.com/news/law-30-an-evolution-toward-on-chain-legal-resolution-in-web3.

strategies of traditional players can stifle the growth of Web3 startups by limiting their ability to compete on cost. This, in turn, creates barriers for new entrants seeking to bring innovative solutions to the market.

To counteract the impact of aggressive pricing by legacy systems, the chapter emphasizes the importance of educating the market on the value proposition of Web3 technologies.[289] Communicating the unique benefits, efficiencies, and decentralization advantages becomes crucial in distinguishing Web3 solutions from the price-centric approach of legacy competitors.

Building Long-Term Sustainability:

While legacy systems may gain a temporary edge through aggressive pricing, the chapter argues that Web3 innovators should focus on building long-term sustainability. Strategies such as emphasizing the intrinsic value of decentralized technologies, fostering community engagement, and differentiating based on principles rather than price become essential.

The chapter explores the potential for collaboration within the Web3 community to resist the short-sighted competition driven solely by price.[290] Collaborative efforts, such as joint marketing campaigns, advocacy for the unique advantages of decentralized technologies, and shared resources, can help the Web3 ecosystem collectively navigate the challenges posed by aggressive pricing strategies.

---

[289] *Id.*

[290] Admin (2023). "Addressing legal concerns in Web3: Legal Compliance in the Decentralized World." KiwiTech at https://www.kiwitech.com/blog/addressing-legal-concerns-in-web3-legal-compliance-in-the-decentralized-world/.

Regulatory frameworks play a crucial role in shaping markets. The chapter briefly touches on the importance of regulatory considerations in addressing unfair pricing practices. Advocacy for a level playing field and fair competition within regulatory frameworks becomes integral for ensuring sustainable growth for Web3 technologies.[291]

The chapter provides insights into strategies that Web3 companies can employ to thrive amidst aggressive pricing from legacy systems. These include diversifying revenue streams, focusing on niche markets, and continuously innovating to stay ahead of the competition.

Conclusion:

In conclusion, the chapter recognizes the challenges posed by market competition fueled by aggressive pricing strategies from legacy systems. By prioritizing education, building long-term sustainability, fostering collaboration, and considering regulatory aspects, the Web3 community can navigate the complexities of the competitive landscape. Rather than succumbing to short-term pressures, Web3 innovators are encouraged to focus on the enduring value and transformative potential of decentralized technologies in the broader market.

---

[291] *Id.*

Chapter 59: Overcoming Resistance to Change - The Business Community and Web3 Adoption

Introduction:

Web3 adoption faces a significant hurdle in the form of resistance to change within the business community. This chapter explores the reasons behind this resistance, its impact on the integration of Web3 technologies, and strategies to overcome such impediments.

The inherent resistance to change is a common challenge across industries, rooted in the comfort of existing processes and fear of the unknown. In the context of Web3 adoption, businesses may be wary of disrupting established norms, especially when transitioning from centralized to decentralized models.

The chapter delves into various factors contributing to the resistance within the business community. These may include concerns about cybersecurity, compliance challenges, the need for significant organizational restructuring, and the uncertainty associated with adopting relatively nascent technologies.

Perceived Risks vs. Potential Benefits:

Business leaders often weigh perceived risks against potential benefits before embracing new technologies.[292] The chapter discusses how the perceived risks of transitioning to Web3, including security vulnerabilities and regulatory uncertainties, may overshadow the transformative benefits, hindering the willingness of businesses to adopt decentralized solutions.

---

[292] Gomez, Jose (2023). "Web3 Websites: The Benefits and Challenges of Adoption." Koombea at https://www.koombea.com/blog/web3-websites/.

To address resistance, the chapter advocates for comprehensive education and awareness initiatives. Businesses need to be equipped with a clear understanding of how Web3 technologies can enhance efficiency, transparency, and security. Educational programs can demystify the complexities associated with the adoption process and alleviate apprehensions.[293]

Demonstrating Real-World Success Stories:

History emphasizes the importance of showcasing real-world success stories from businesses that have successfully integrated Web3 technologies. Highlighting instances where the transition led to increased efficiency, cost savings, and improved customer trust can serve as powerful motivators for businesses contemplating adoption.[294]

Collaboration with early adopters within the business community can help overcome resistance. By forming partnerships with organizations that have successfully integrated Web3 solutions, businesses can gain insights, share best practices, and mitigate concerns related to implementation challenges.

The regulatory landscape surrounding Web3 technologies is evolving. The chapter discusses the significance of addressing regulatory concerns to alleviate uncertainties for businesses. Engaging with regulatory bodies, advocating for clear guidelines, and actively participating in the shaping of regulations can

---

[293] Staff (2023). "Web3: incredible Advantages and Unexpected Risks." SelfKey at https://selfkey.org/web3-incredible-advantages-and-unexpected-risks/.

[294] Walsh, Joe & Rivet, Emmanuelle (2023). "Trust can be the foundation of Web3 success: Here's how to start building it." PWC at https://www.pwc.com/us/en/tech-effect/emerging-tech/building-trust-into-web3.html.

contribute to a more conducive environment for Web3 adoption.[295]

Businesses resistant to change may be motivated by incentives. The chapter explores how governments, industry associations, and advocacy groups can play a role in incentivizing innovation. Offering grants, tax incentives, or recognition for businesses embracing Web3 technologies can foster a more innovation-friendly ecosystem.[296]

The commitment and vision of leadership are pivotal in overcoming resistance. The chapter emphasizes the role of leaders in championing the cause of Web3 adoption, fostering a culture of innovation, and aligning the organization's goals with the transformative potential of decentralized technologies.

Conclusion:

In conclusion, the chapter recognizes the formidable challenge of resistance to change within the business community as a roadblock to Web3 adoption. By addressing concerns through education, showcasing success stories, fostering collaboration, tackling regulatory uncertainties, providing incentives, and leveraging strong leadership, the business community can overcome resistance and unlock the transformative potential of Web3 technologies. The journey toward a decentralized future requires collective efforts to break the chains of resistance and usher in a new era of innovation and efficiency.

---

295 *Id.*

296 Gomez (2023) *supra*.

Chapter 60: Scaling the Heights - The Crucial Challenge of Scalability in Web3

Introduction:

Scalability stands as a pivotal challenge in the realm of Web3, where ambitions of decentralization and widespread adoption collide with the practical constraints of technology. This chapter explores why scalability is a critical challenge for Web3, its impact on performance, and potential strategies to overcome this hurdle.

> *Scalability: A systems ability to handle an increase in the amount of work or growth.*

Scalability refers to a system's ability to handle an increasing amount of work or growth efficiently. In the context of Web3, which envisions a decentralized and democratized digital landscape, scalability becomes a bulwark. It determines how well the system can accommodate a growing user base, increasing transaction volumes, and expanding functionalities without compromising performance.

Analysis of the repercussions of poor scalability in Web3 ecosystems is vital to understanding why scalability is a fundamental factor in any system. Inefficiencies in handling a surge in transactions or accommodating a growing user base can lead to network congestion, slower transaction processing times, increased fees, and a degraded user experience. These issues threaten the very essence of decentralized systems, which aim to provide seamless, efficient, and inclusive digital interactions.

Web3 platforms, often built on blockchain technology, face inherent challenges related to scalability. The decentralized nature of these networks, while offering security and transparency, presents obstacles like limited transaction throughput, slower confirmation times, and increased energy consumption. These challenges become pronounced as the demand for decentralized applications (DApps) and blockchain-based services grows.

Increasing Transaction Throughput:

> *Throughput: the amount or quality of data passing through a system or process.*

The chapter explores strategies to enhance transaction throughput, a key facet of scalability. Layer 2 scaling solutions, such as sidechains and state channels, are discussed as mechanisms to offload transactions from the main blockchain, thus increasing the overall capacity.[297] Additionally, advancements in consensus algorithms, like proof-of-stake (PoS) and sharding, aim to distribute transaction processing across the network more efficiently.

Balancing Security and Scalability:

Scalability efforts must navigate the delicate balance between efficiency and security. The chapter emphasizes the importance of maintaining the robust security features inherent in decentralized systems while pursuing scalability solutions.[298] Innovations like optimistic rollups and zk-rollups are explored for

---

[297] Staff (2023). "Web3 Scalability Solutions: Tackling the Blockchain Bottleneck." JumboBlockchain at https://jumbochain.medium.com/web3-scalability-solutions-tackling-the-blockchain-bottleneck-77671c0c64bf.

[298] Adams, Jordan (2023). "What are the Scalability issues of Web3? Blockchain Scalability." Doubloin at https://www.doubloin.com/learn/scalability-issues-web3.

their potential to enhance efficiency without compromising the security and integrity of the underlying blockchain.

Interoperability is positioned as a key enabler of scalability. The chapter discusses the significance of creating interoperable frameworks that allow different blockchain networks to communicate seamlessly. Such interoperability not only enhances the overall scalability of the Web3 ecosystem but also fosters a more interconnected and collaborative decentralized landscape.

Innovation emerges as a driving force in scalability solutions. The chapter explores how ongoing research and development are crucial for identifying novel approaches to scalability. From the exploration of new consensus mechanisms to the evolution of blockchain architectures, constant innovation is essential to surmount the challenges posed by scalability.[299]

Conclusion:

In conclusion, the chapter recognizes scalability as a critical challenge that necessitates strategic solutions for the sustained success of Web3. As the demand for decentralized technologies continues to grow, the pursuit of scalability becomes a journey of innovation, compromise, and careful calibration. By addressing the current challenges, prioritizing transaction throughput, maintaining a balance between security and efficiency, fostering interoperability, and embracing continuous innovation, the Web3 community can scale new heights and fulfill its promise of a decentralized, scalable, and inclusive digital future.

[299] Smithmyer, Christopher (2023). Digital Exchequer. Elite Exclusivity: Detroit.

Conclusion Part VI

In the voyage through Part VI, we embarked on a journey delving into the challenges of Web3, uncovering the myriad of possibilities that define its landscape. As we navigate the uncharted territories of regulatory compliance, security concerns, talent acquisition, and the clash with legacy systems, the conclusion of this segment invites reflection on the collective ecosystem woven by these chapters.

Web3, with its decentralized aspirations, encounters the formidable barrier of regulatory compliance. The chapters explored the intricacies of complying with diverse and evolving regulations. The conclusion drawn is that, while regulatory frameworks are essential for stability and security, a harmonized, globally accepted set of standards is imperative to foster innovation and unlock the full potential of Web3.

Security emerged as a central theme, and rightly so, as Web3 strives to build trust in decentralized ecosystems. The discussion navigated through the vulnerabilities and potential mitigations, emphasizing the need for a proactive approach. The conclusion underscores that robust security measures, coupled with an agile response to emerging threats, are prerequisites for sustaining user confidence and the integrity of Web3 platforms.

Chapter 53 underscored the pivotal role of talent acquisition and training in nurturing a vibrant and innovative Web3 ecosystem. As organizations grapple with the demand for blockchain and Web3 expertise, the conclusion echoes the importance of a talent pool, both within and beyond organizational boundaries, to drive technological advancements and shape the future of decentralized technologies.

The narrative explored the challenges associated with aligning technological infrastructure with the demands of a rapidly evolving digital universe. The conclusion asserts that overcoming this challenge requires a holistic approach – one that involves collaboration between public and private entities, strategic investments in technological upgrades, and a willingness to embrace transformative shifts in infrastructure design.

Chapters 55 and 56 scrutinized the complexities of integrating Web3 technologies with legacy systems. The conclusion emphasizes the critical need for a phased, adaptive strategy. Leveraging artificial intelligence (AI) as a tool to smooth the integration process, the Web3 community must navigate the terrain carefully, ensuring that technological evolution aligns seamlessly with legacy systems without compromising efficiency or security.

Resistance to change, exacerbated by fear-driven narratives in mainstream media, was dissected in Chapter 57. The chapters collectively underscore that dispelling myths and fostering understanding through educational initiatives are essential to quell apprehensions and enable the broader acceptance of Web3 technologies.

Chapters 58 and 59 unraveled the challenges posed by market s and business communities' resistance to change. The conclusion posits that fostering an environment of collaboration, driving awareness, and demonstrating the tangible benefits of Web3 technologies are vital for overcoming these hurdles.

The journey culminated with a deep dive into the perpetual challenge of scalability in Chapter 60. The conclusion underscores that the pursuit of scalability is not a singular endeavor but an ongoing process of innovation, adaptation, and finding the delicate equilibrium between efficiency and security.

In conclusion, Part VI serves as a compass, guiding stakeholders through the complexities of Web3. It illuminates the challenges, unveils potential solutions, and echoes the collective call for collaboration, innovation, and adaptability. As we navigate the landscapes of regulation, security, talent, infrastructure, legacy integration, societal perception, markets, and scalability, the true essence lies in recognizing these challenges not as roadblocks but as catalysts for growth and evolution in the realm of Web3. The journey continues, each chapter contributing to the collective wisdom that will shape the future of decentralized technologies.

Part VII: Building the Fintech Ecosystem-Developing Infrastructure

Introduction Part VII

Welcome to the seventh segment of our exploration into the vast and transformative landscape of Web3. In Part VII, we embark on a journey that transcends the technical modalities, focusing on the fundamental elements that lay the foundation for a thriving Web3 ecosystem within nations. Here, we delve into the realms of regulation, education, talent acquisition, and infrastructure development, unraveling the threads that, when woven together, create an ecosystem of innovation and progress.

The chapters in this section cast a spotlight on the paramount importance of regulatory clarity in fostering a conducive environment for Web3 technologies. In Chapter 61, we contemplate the necessity of clear and flexible regulations for optimizing the potential of Web3 on a global scale. As the digital landscape evolves, the adaptability of regulations becomes a bulwark, ensuring that innovation can flourish within a secure and regulated framework.

Chapter 62 introduces the concept of regulatory sandboxes—a space for controlled experimentation and learning. We explore how these sandboxes become incubators for innovative solutions, offering businesses a platform to test their Web3 initiatives without the constraints of rigid regulations. The sandbox approach is more than a regulatory tool; it is a testament to the collaborative spirit between regulators and businesses in fostering innovation.

In Chapter 63, the narrative shifts towards the collaborative engagement between Web3 projects and regulatory bodies. We delve into the symbiotic relationship needed for the harmonious coexistence of innovation and governance. As businesses and regulators join forces, a dialogue ensues, ensuring that the regulatory framework aligns with the dynamism of the Web3 landscape, striking a delicate balance between progress and security.

Education emerges as a fundamental part in Chapters 64 and 65, where we explore the pivotal role of regional cryptocurrency education programs and global talent attraction initiatives. These chapters underscore the importance of nurturing a knowledgeable and skilled workforce to propel nations into the forefront of the Web3 revolution. Education becomes the catalyst for unlocking the full potential of these transformative technologies.

Chapters 66 to 69 venture into the critical domains of talent acquisition, infrastructure development, and financial inclusion. We dissect the anatomy of a robust digital infrastructure and its role in attracting global talent. Financial inclusion becomes a bulwark, ensuring that the benefits of Web3 technologies are accessible to all, creating a truly inclusive and sustainable ecosystem.

Chapter 68 explores the importance of networking programs, startup launch systems, and collaboration hubs. These elements are the connective tissue of a vibrant Web3 ecosystem, nurturing a community of innovators. Chapters 69 and 70 confront challenges such as resistance to change and scalability concerns, presenting opportunities for nations to position themselves as pioneers in overcoming these obstacles.

As we delve into Part VII, we invite you to explore the multifaceted aspects of creating an environment where Web3 technologies not only survive but thrive. The chapters ahead unravel the dance between regulation, education, infrastructure, and collaboration—key elements in sculpting a future where the full potential of Web3 is realized on a global scale.

Chapter 61: Clear and Flexible Regulations (Global)

Introduction:

In the ever-expanding realm of Web3, where decentralized technologies redefine industries and reshape economies, the importance of regulatory frameworks cannot be overstated. As we navigate this landscape, one truth becomes abundantly clear: for Web3 to reach its optimal potential on a global scale, a delicate balance of clarity and flexibility in regulatory structures is essential.

Web3 technologies, including blockchain, smart contracts, and decentralized applications, traverse national borders seamlessly. Yet, the absence of standardized global regulations has given rise to a fragmented landscape.[300] Regulatory frameworks vary significantly from one jurisdiction to another, creating challenges for businesses, developers, and users attempting to navigate this jigsaw of paths, dead ends, and dangerous terrain.

At the heart of effective regulation lies clarity – a fundamental prerequisite for fostering innovation and ensuring compliance. Clear guidelines empower businesses and individuals to understand the rules of engagement, facilitating the development and deployment of Web3 technologies with confidence. Clarity diminishes uncertainty, offering a stable foundation for growth and fostering a regulatory environment where all stakeholders can thrive.

Global Interconnectedness:

Web3's global nature demands a regulatory approach that acknowledges and accommodates its interconnectedness. The

---

[300] Mulligan, Cathy (2018). "Blockchain and Sustainable Growth." UN Chronicle at https://www.un.org/en/un-chronicle/blockchain-and-sustainable-growth.

decentralized nature of blockchain, for instance, means that transactions and activities can originate from and impact multiple jurisdictions simultaneously.[301] A nuanced regulatory framework, recognizing the borderless nature of Web3 technologies, is paramount for their seamless integration into the global economic fabric.

While clarity is imperative, the nature of Web3 technologies calls for regulatory frameworks that are flexible and adaptive. The pace of technological evolution within the Web3 space is rapid, and static regulations risk becoming obsolete before the ink dries. Flexibility allows regulators to respond to emerging challenges and opportunities, fostering an environment where innovation can flourish without being stifled by outdated rules.

Clear and comprehensible regulations serve as incentives for compliance. Businesses and individuals are more likely to adhere to rules when they understand the reasons behind them. Clarity fosters a culture of compliance, steering the Web3 community toward responsible and ethical practices. In turn, this engenders trust, a crucial element for the broader acceptance of decentralized technologies.[302]

Balancing Innovation and Consumer Protection:

The regulatory tightrope extends to striking a delicate balance between fostering innovation and ensuring consumer protection. Clarity in guidelines allows innovators to explore new frontiers within the regulatory boundaries, providing the necessary room for experimentation.[303] Simultaneously, clear

---

[301] Self, Karen (2023). "How Blockchain is Affecting Globalization." DIVA at https://www.diva-portal.org/smash/get/diva2:1481911/FULLTEXT01.pdf.
[302] *Id.*
[303] Staff (2023). "Blockchain for Protecting Consumer Rights." Reintech at https://reintech.io/blog/blockchain-protecting-consumer-rights.

regulations act as a safeguard, protecting consumers from potential risks associated with rapidly evolving technologies.

In the absence of uniform global regulations, regulatory arbitrage becomes a concern. Businesses might seek jurisdictions with lax regulations, potentially leading to uneven landscapes and regulatory competition. A harmonized global approach, emphasizing clarity and flexibility, can mitigate these challenges, promoting fairness and consistency in the application of rules.

Conclusion:

As Web3 technologies continue to redefine our digital landscape, the call for clear and flexible regulatory frameworks echoes louder than ever. The harmonization of global regulations, emphasizing clarity and adaptability, emerges as a fundamental element for the widespread and responsible integration of Web3 into our interconnected world. Striking the delicate balance between fostering innovation and safeguarding the interests of users, clear and flexible regulations provide the navigational aids required to chart a course toward a decentralized future that benefits all.

Chapter 62: Regulatory Sandboxes in Web3 - Fostering Innovation through Controlled Experimentation (Regional)

Introduction:

In the frontier of Web3 technologies, where innovation outpaces regulatory frameworks, the concept of regulatory sandboxes has emerged as a crucial tool. These controlled environments provide a space for experimentation, enabling regulators, businesses, and developers to collaboratively navigate the uncharted territories of decentralized technologies. This chapter explores the importance of regulatory sandboxes in shaping effective and responsive regulations for the Web3 era.

A regulatory sandbox is essentially a controlled environment where businesses and developers can test innovative products, services, and business models without immediately incurring all the normal regulatory consequences.[304] It allows stakeholders to explore the possibilities and implications of new technologies, providing regulators with valuable insights to frame appropriate regulations.

One of the primary roles of regulatory sandboxes is to foster innovation within the Web3 space. As technologies like blockchain, smart contracts, and decentralized applications advance at an unprecedented pace, traditional regulatory frameworks struggle to keep up. Regulatory sandboxes act as incubators, allowing innovators to experiment, iterate, and fine-tune their solutions, ultimately contributing to the broader ecosystem's growth.

---

[304] Staff (2023). "Regulation for Inclusive Digital Finance." CGAP at https://www.cgap.org/topics/collections/regulatory-sandboxes.

Regulatory sandboxes promote a collaborative learning environment where regulators actively engage with businesses and developers. This two-way dialogue ensures that regulators gain a deep understanding of the technologies being tested, the associated risks, and potential benefits. Simultaneously, businesses receive valuable feedback, helping them align their innovations with regulatory expectations.

Adapting to Technologies:

Web3 technologies are characterized by their and ever-evolving nature. Regulatory sandboxes provide a space where regulators can adapt and refine rules in real-time based on the insights gained from ongoing experiments. This agility is crucial in an environment where the regulatory landscape needs to keep pace with the rapid evolution of decentralized technologies.[305]

While fostering innovation is a primary goal, regulatory sandboxes also play a crucial role in understanding and mitigating risks associated with emerging technologies. The controlled environment allows regulators to assess potential risks without hindering the broader development of the Web3 ecosystem.[306] This delicate balance between encouraging innovation and safeguarding against risks is vital for responsible technological advancement.

Regulating cutting-edge technologies requires a deep understanding of their intricacies. Regulatory sandboxes provide regulators with hands-on experience, allowing them to build expertise in the nuances of Web3 technologies. This expertise

---

[305] *Id.*

[306] Simpson, Erin & Conner, Adam (2021). "How to Regulate Tech: A Technology Policy Framework for Online Services." American Progress at https://www.americanprogress.org/article/how-to-regulate-tech-a-technology-policy-framework-for-online-services/.

becomes invaluable when crafting regulations that strike the right balance between encouraging innovation and protecting stakeholders.

Web3 technologies often transcend traditional sector boundaries. Regulatory sandboxes facilitate cross-sector collaboration by bringing together stakeholders from finance, technology, healthcare, and more.[307] This interdisciplinary approach ensures that regulations are holistic and consider the diverse applications of decentralized technologies across various industries.

Transparent and collaborative experimentation in regulatory sandboxes enhances public trust in the regulatory process. By actively involving stakeholders in the development of regulations, regulators can demonstrate a commitment to fairness, accountability, and responsiveness. This, in turn, fosters confidence among businesses, developers, and the general public in the regulatory framework governing Web3 technologies.

Conclusion:

In the landscape of Web3, regulatory sandboxes emerge as crucibles for innovation, collaboration, and regulatory evolution. By providing a controlled environment for experimentation, these sandboxes empower regulators to adapt to the rapidly changing technological landscape while fostering the responsible development of decentralized technologies. As the Web3 era unfolds, the significance of regulatory sandboxes cannot be overstated, offering a glimpse into a future where regulations are not barriers but enablers of innovation.

---

[307] *Id.*

Chapter 63: Collaborating with Regulators - Navigating Web3 Projects through Regulatory Partnerships (Global)

Introduction:

In the interplay between innovation and regulation within the Web3 landscape, collaboration with regulators emerges as a crucial element. As decentralized technologies redefine industries and challenge traditional regulatory frameworks, close cooperation between Web3 projects and regulatory bodies becomes imperative. This chapter delves into the importance of fostering collaborations with regulators to navigate the complex regulatory environment surrounding Web3 initiatives.

Web3 projects often operate in an environment characterized by regulatory uncertainty. Engaging in open and transparent collaboration with regulators enables projects to navigate this uncertainty more effectively.[308] By establishing lines of communication, projects can gain insights into regulatory expectations and contribute to the development of frameworks that foster innovation while ensuring compliance.

Building Trust and Credibility:

Collaborating with regulators builds trust and credibility for Web3 projects. The decentralized nature of blockchain and other Web3 technologies may, at times, be perceived with

---

[308] Staff (2023). Astra May 2023 Recap: Accelerating Web3 Regulation with Progressive Partnerships and Decentralized Compliance Solutions." Astra at https://astraprotocol.medium.com/astra-may-2023-recap-accelerating-web3-regulation-with-progressive-partnerships-and-decentralized-63f43c719169.

skepticism by regulatory bodies.[309] Proactive engagement, where projects willingly share their technologies, methodologies, and intentions, helps demystify complex innovations, fostering a sense of trust among regulators.

Many regulators may not have an in-depth understanding of the intricacies of Web3 technologies. Collaborative efforts allow Web3 projects to take on an educational role, providing regulators with the necessary knowledge to make informed decisions.[310] This educational exchange is a two-way street, as it also enables projects to understand regulatory concerns and expectations.

Regulations First Mindset: The idea that governments know better than businesses which leads to failed policy being created.

Web3 projects can actively contribute to the shaping of regulatory frameworks by collaborating with regulators. Sharing experiences, best practices, and participating in consultations enable projects to have a direct impact on the development of regulations. This collaborative approach helps ensure that regulations align with the realities and goals of the Web3 ecosystem.

A regulatory-first mindset involves considering regulatory implications from the inception of a Web3 project. By involving regulators early in the development process, projects can

---

309 Staff (2023). "Web3 Startups and the Power of Trust: Building Strong Online Communities." TDX at https://medium.com/@TDXbiz/web3-startups-and-the-power-of-trust-building-strong-online-communities-0d58c8155671.

310 Hymel, Cory (2023). "Whey is web3 education in government critical to adoption?" Gigster at https://gigster.com/blog/web3-education-government/.

address potential concerns, design features with compliance in mind, and reduce the likelihood of facing regulatory hurdles later on. This proactive approach reflects a commitment to responsible innovation.

Creating a Supportive Ecosystem:

Regulatory collaboration extends beyond individual projects and contributes to the creation of a supportive ecosystem. Web3 projects can work collectively with regulators to establish industry standards, share information on emerging trends, and collectively address challenges.[311] This collaborative ecosystem benefits all stakeholders, fostering an environment where innovation and compliance coexist.

Collaborating with regulators enables Web3 projects to proactively address compliance requirements. Rather than reacting to enforcement actions, projects that maintain an ongoing dialogue with regulators can anticipate and align with evolving compliance expectations. This approach minimizes legal risks and positions projects as responsible contributors to the regulatory landscape.

Long-term viability for Web3 projects hinges on their ability to operate within legal and regulatory frameworks. Collaborating with regulators ensures that projects remain adaptable to changing regulatory landscapes, reducing the risk of facing legal challenges that could jeopardize their existence. This collaborative stance is a strategic investment in the sustainability of Web3 initiatives.

Conclusion:

In the ecosystem of Web3 innovation, collaboration with regulators stands out as a guiding thread. By actively engaging with regulatory bodies, Web3 projects not only navigate current

---

[311] Smithmyer (2023a), *supra*.

regulatory landscapes but also contribute to the development of frameworks that foster innovation responsibly. The symbiotic relationship between Web3 projects and regulators holds the promise of a future where cutting-edge technologies coexist harmoniously with regulatory considerations.

Chapter 64: Empowering Regions through Cryptocurrency Education for Web3 Development (Regions)

Introduction:

In the landscape of Web3 development, the importance of cryptocurrency education at the regional level cannot be overstated. As blockchain technology and cryptocurrencies redefine traditional business models, fostering a comprehensive understanding of these innovations becomes an important element for regional growth. This chapter explores the multifaceted significance of cryptocurrency education in propelling the development of Web3 businesses across diverse regions.

Educational Foundations for Web3:

Web3, built on decentralized technologies like blockchain, introduces novel concepts that require a foundational understanding of cryptocurrency.[312] At the heart of regional Web3 development lies the need to equip individuals, businesses, and communities with the knowledge to navigate and leverage these transformative technologies effectively.

Cryptocurrency education serves as a democratizing force, opening doors to opportunities that were once limited by traditional financial systems.[313] Regions that prioritize comprehensive cryptocurrency education empower their residents to participate in the global digital economy, fostering financial inclusion and reducing economic disparities.

---

[312] Alchemy at https://www.alchemy.com/best/web3-education-resources.

[313] Notomoro (2023). "How Web3 in EdTEch will bring Revolution in Education." Webisoft at https://webisoft.com/articles/web3-in-edtech/ (Use the system to promote the system).

Regions that invest in cryptocurrency education often witness the emergence of innovation hubs. These hubs become focal points for collaboration, research, and the development of Web3 projects. By nurturing a pool of skilled individuals with a deep understanding of cryptocurrencies, regions position themselves as hubs for cutting-edge technological advancements.

Cryptocurrency education lays the groundwork for a new wave of entrepreneurship. Individuals equipped with the knowledge of blockchain and cryptocurrencies are more likely to venture into Web3 startups and contribute to the regional ecosystem.[314] This influx of entrepreneurial activity has the potential to drive economic growth and job creation.

Attracting Investment:

Regions with a well-educated workforce in cryptocurrency and Web3 technologies become attractive destinations for investment. Investors seek environments where they can find skilled talent and a supportive ecosystem for blockchain-based ventures.[315] Cryptocurrency education acts as a magnet, drawing in investment that further accelerates regional development.

Beyond the intricacies of blockchain technology, cryptocurrency education enhances financial literacy at the regional level. Individuals gain insights into managing digital assets, understanding market's, and making informed financial

---

[314] Gilmore, Aimee (2023). "Into the Verse-"Web3 and Education: Transforming Learning through Decentralized Systems." Into The Verse at https://medium.com/intotheverse-journal/intotheverse-web3-and-education-transforming-learning-through-decentralized-systems-461f08a94c8d.

[315] Chimelu, Gloria (2023). "How to invest in Web 3.0." Investopedia at https://www.investopedia.com/how-to-invest-in-web-3-0-7480982.

decisions. This elevated financial literacy contributes to overall economic resilience within the region.[316]

Cryptocurrency education equips regional stakeholders, including policymakers and regulators, with the knowledge needed to address regulatory challenges. Informed regulatory frameworks are crucial for fostering innovation while safeguarding against potential risks. Regions that proactively engage in cryptocurrency education are better positioned to create balanced and effective regulatory environments.[317]

An essential aspect of cryptocurrency education is instilling a sense of ethical responsibility within the community. As regions embrace Web3 technologies, promoting ethical practices becomes paramount. Education empowers individuals to understand the ethical considerations surrounding cryptocurrencies, promoting responsible innovation.

Cryptocurrency education fosters a sense of community and collaboration within regions. Communities of like-minded individuals, businesses, and organizations emerge, creating networks that share knowledge, resources, and support. These communities become pillars of regional Web3 development, encouraging collective progress.

Conclusion:

The development of Web3 businesses at the regional level is tied to the depth of cryptocurrency education. Regions that recognize the transformative potential of blockchain technologies and invest in comprehensive education initiatives position themselves at the forefront of the global Web3 landscape. The journey toward regional empowerment through

---

[316] Kamal, Abu Hena Mostofa (2024). How to invest in Web3: Know everything before you invest." BD Task at https://www.bdtask.com/blog/how-to-invest-in-web3.

[317] Chimelu, *supra*.

cryptocurrency education is not merely a technological evolution but a societal transformation that propels communities into the digital future.

Chapter 65: Attracting Global Talent for Web3 Development (Regional)

Introduction

In the competitive realm of Web3 development, attracting global talent is a pivotal aspect that can catalyze innovation and drive sustained growth. As the landscape of decentralized technologies evolves, the demand for skilled individuals versed in Web3 frameworks becomes increasingly pronounced. This chapter delves into the strategies and considerations involved in attracting top-tier talent at a global level for Web3 initiatives.

Attracting global talent for Web3 development requires a strategic approach. Companies and regions aspiring to lead in the Web3 space must actively seek out individuals with diverse skills, experiences, and perspectives.[318] A strategic talent acquisition plan involves understanding the specific needs of Web3 projects and identifying individuals who can contribute meaningfully.

The success of Web3 initiatives hinges on diversity and inclusion. Global talent acquisition should prioritize creating diverse teams that bring varied insights to the table. Embracing individuals from different backgrounds fosters creativity, innovation, and a richer ecosystem of ideas, essential for tackling the multifaceted challenges of Web3 development.

Creating an Enabling Ecosystem:

Top-tier talent is drawn to environments that nurture innovation and provide opportunities for professional growth. Companies and regions aiming to attract global talent for Web3 should

---

[318] Staff (2023). "Recruiting for Blockchain: The challenges and how to overcome them." Crypto Recruit at https://www.cryptorecruit.com/news/recruiting-for-blockchain-the-challenges-and-how-to-overcome-them/.

focus on creating an enabling ecosystem. This includes offering mentorship programs, continuous learning opportunities, and a supportive culture that encourages experimentation and risk-taking.

In the global talent marketplace, compensation and benefits play a crucial role. To attract the best minds in Web3 development, companies must offer competitive packages that reflect the value of the skills and expertise required. Beyond monetary compensation, comprehensive benefits packages, including health and wellness initiatives, contribute to a compelling value proposition.

Web3 development often transcends geographical boundaries, making remote work an attractive option for global talent. Embracing remote work opportunities expands the talent pool and enables companies to tap into expertise from around the world. This flexibility is particularly appealing to skilled professionals seeking a work-life balance aligned with their preferences.

Showcasing Impactful Projects:

Top talent seeks projects that are not only technically challenging but also have a meaningful impact. Companies should showcase their ongoing and prospective Web3 projects, emphasizing how these initiatives contribute to broader societal goals. Demonstrating a commitment to projects with real-world implications can be a significant factor in attracting global talent.

Collaboration and Networking:

Active participation in global communities, conferences, and networking events is instrumental in talent acquisition. Companies involved in Web3 development should engage with the wider ecosystem, fostering relationships with potential candidates. Building a strong presence in relevant communities helps create a network of professionals who are aware of and interested in contributing to Web3 initiatives.[319]

Conclusion

In conclusion, attracting global talent for Web3 development is a nuanced undertaking that demands a holistic approach. Companies and regions that position themselves as attractive hubs for top-tier talent prioritize diversity, create enabling ecosystems, offer competitive compensation, embrace remote work, showcase impactful projects, and actively engage in collaboration and networking. As the Web3 landscape continues to evolve, the ability to attract and retain global talent emerges as a defining factor in the success of transformative initiatives. By strategically investing in talent acquisition, companies and regions can propel themselves to the forefront of Web3 innovation.

[319] Web3 (2023). talent at https://web3-talents.io/.

Chapter 66: Robust Digital Infrastructure (National)

Introduction

In the rapidly evolving landscape of blockchain, Web3, and cryptocurrency, the need for a robust digital infrastructure is foundational to drive innovation, ensure security, and enable widespread adoption. This chapter explores the critical components and imperatives of establishing a resilient digital foundation for the decentralized technologies shaping the future.

Digital Infrastructure: The electric grid, internet system, free speech rights, and education system allowing for the development of Web3 ecosystems.

A robust digital infrastructure encompasses the underlying technological framework supporting blockchain, Web3, and cryptocurrency ecosystems. This foundation includes decentralized networks, consensus mechanisms, cryptographic protocols, and distributed ledgers. Each element plays a pivotal role in ensuring the reliability, security, and scalability of the entire system.

One of the primary challenges in blockchain and Web3 development is achieving scalability without compromising performance. A robust digital infrastructure must facilitate seamless scaling to accommodate a growing user base and increased transaction volumes. Solutions such as layer 2 scaling, sharding, and optimized consensus algorithms contribute to enhanced performance and scalability.

Interoperability Across Networks:

Interoperability is a key requirement for a cohesive digital infrastructure. Blockchain and Web3 projects often operate on diverse networks and protocols. Establishing seamless interoperability enables the frictionless exchange of assets, data, and functionalities across different platforms, fostering a more interconnected and versatile ecosystem.[320]

Security is paramount in the digital realm, particularly in decentralized ecosystems where trust is distributed.[321] A robust infrastructure incorporates state-of-the-art cryptographic techniques, secure consensus mechanisms, and rigorous validation processes. Resilience against cyber threats, including 51% attacks and vulnerabilities, is integral to maintaining the integrity of the entire network.

Decentralized storage and computation mechanisms contribute to the overall efficiency of the infrastructure.[322] Utilizing distributed storage solutions ensures data redundancy and reduces the risk of single points of failure. Decentralized computation, often facilitated by smart contracts, enables the execution of complex algorithms without relying on centralized servers.

A well-designed digital infrastructure includes robust governance mechanisms to manage protocol upgrades and modifications. Decentralized autonomous organizations (DAOs) and community-driven governance models empower stakeholders to participate in decision-making processes.[323] This

---

[320] Staff (2023). "Web3 Interoperability: Bridging the Future of the Internet." Giakaaweb3 at https://medium.com/@sm_28205/web3-interoperability-bridging-the-future-of-the-internet-a38370507cd3.

[321] Smithmyer (2023a), *Supra*.

[322] Giakaaweb3, *supra*.

[323] Smithmyer (2023a), *Supra*.

inclusivity enhances the adaptability and upgradability of the infrastructure.

User-Friendly Interfaces:

For widespread adoption, user-friendly interfaces are crucial. The digital infrastructure should support intuitive applications and wallets that simplify interactions for both novice and experienced users.[324] Improving the user experience fosters broader accessibility, contributing to the democratization of blockchain, Web3, and cryptocurrency technologies.

Regulatory Compliance and Privacy:

A comprehensive digital infrastructure acknowledges the importance of regulatory compliance and user privacy. Implementing privacy-preserving technologies, adhering to regulatory standards, and incorporating identity management solutions contribute to a secure and legally compliant environment. Striking the right balance between transparency and privacy is vital for sustainable growth.

As the environmental impact of blockchain and cryptocurrency becomes a focal point, a robust digital infrastructure considers sustainability.[325] Implementing energy-efficient consensus algorithms and exploring eco-friendly alternatives align with the broader commitment to environmental responsibility.

Conclusion:

In conclusion, building a robust digital infrastructure for blockchain, Web3, and cryptocurrency is a multifaceted

---

[324] *See* "Why do I need to know about UI and UX Design in Web3?" at https://www.linkedin.com/pulse/why-do-i-need-know-ui-ux-design-web3-neue-world/.

[325] Staff (2023). "Regulatory Compliance in Web3: A Guide for business." Entertainment Law Miami at https://entertainmentlawyermiami.com/regulatory-compliance-in-web3-a-guide-for-businesses/.

endeavor that demands careful consideration of scalability, interoperability, security, decentralized storage, governance, user interfaces, regulatory compliance, and environmental sustainability. This foundational work lays the groundwork for the continued evolution and widespread adoption of decentralized technologies, ensuring they become integral components of the digital landscape. By addressing these critical aspects, the digital infrastructure becomes a resilient and adaptive backbone, supporting the transformative potential of blockchain, Web3, and cryptocurrency.

Chapter 67: Attracting Web3 Businesses: The Power of Incentives and Financial Support (National).

Introduction:

In the chaotic realm of Web3 technologies, the competition to attract innovative businesses is fierce. Regions and countries keen on positioning themselves as hubs for blockchain and decentralized technologies recognize the pivotal role of incentives and financial support. This chapter delves into the compelling impact of strategic incentives and financial backing in drawing Web3 businesses to a particular locale.

Incentives serve as powerful magnets, influencing the decisions of businesses seeking favorable environments for growth and innovation.[326] Regions that offer incentives tailored to the unique needs of Web3 enterprises create an attractive proposition. These incentives can take various forms, including tax breaks, grants, regulatory support, and collaborative initiatives with local institutions.

Financial Support

Financial support acts as a catalyst for Web3 businesses, providing the necessary resources for development, research, and market expansion. Governments and municipalities that allocate funds specifically for blockchain and decentralized technology projects demonstrate a commitment to fostering a thriving ecosystem. Financial support can manifest through grants, subsidies, venture capital initiatives, and access to low-interest loans.

---

[326] Paul, Christina (2023). "Top 10 Countries to Invest inWeb3 Marketing Services." Coin Monks at https://medium.com/coinmonks/top-10-countries-to-invest-in-web-3-marketing-service-4cf06ad4105a.

Governments can stimulate innovation by introducing tax incentives that directly benefit Web3 businesses. Reductions in corporate taxes, research and development (R&D) tax credits, and exemptions for certain blockchain-related activities create an environment where businesses can allocate more resources to innovation and technological advancement.[327]

Direct financial support in the form of grants and subsidies accelerates the growth of Web3 startups and enterprises. Governments or regional bodies can establish funds dedicated to supporting blockchain and decentralized technology initiatives.[328] These funds can be allocated through competitive grant programs, providing essential capital for projects with high potential.

The establishment of regulatory sandboxes is a forward-thinking approach to fostering Web3 innovation. These controlled environments allow businesses to experiment with new technologies within a framework that offers flexibility and support. Regulatory sandboxes provide a space for collaboration between businesses and regulatory bodies, fostering a deeper understanding of the unique challenges and opportunities in the Web3 space.

---

[327] Ganguly, Aditi (2023). "The Metaverse Tax Conundrum: How it Could become the Ultimate Tax Haven." Yahoo Finance at https://finance.yahoo.com/news/metaverse-tax-conundrum-could-become-195219420.html?guccounter=1&guce_referrer=aHR0cHM6Ly93d3cuZ29vZ2xlLmNvbS8&guce_referrer_sig=AQAAAKsYlokSVxO2cGcgxw4rVn2IOcX9cDuN6ebczUnqWDmPNEEu-9uOlxGhRUvhHzW-D80K7vW7MqL3FtVr-v0RevwyJ4iy6o4pIAvhXgNs17aamw8jFuTkef_qvJeQ_beJ2Fme_n__R6EQo7JKJAUlnlQPtuQxc5-tMAcMlMXHJ1ne.

[328] Web3 Foundation Grants at https://grants.web3.foundation/.

Collaborative Ecosystems:

Financial support can extend beyond direct funding to the creation of collaborative ecosystems. Regions that actively facilitate partnerships between Web3 businesses, research institutions, and traditional industries nurture a holistic environment for innovation.[329] Collaborative spaces, incubators, and accelerators can further enhance the vibrancy of the Web3 ecosystem.

Access to Skilled Talent:

Financial support can be directed towards developing a skilled talent pool that aligns with the needs of Web3 businesses. Initiatives such as educational programs, training workshops, and partnerships with academic institutions contribute to the availability of a workforce well-versed in blockchain and decentralized technologies.

Highlighting success stories within the Web3 ecosystem can attract further attention and investment. Regions that actively showcase the achievements of local Web3 businesses demonstrate a commitment to the industry's growth. Success stories serve as inspiration for aspiring entrepreneurs and reinforce the region's credibility as a hub for innovation.

Conclusion:

In conclusion, the combination of incentives and financial support creates a compelling proposition for regions aspiring to become centers of Web3 innovation. Governments and municipalities that strategically design incentives, offer financial backing, and cultivate collaborative ecosystems position themselves as attractive destinations for Web3 businesses. The ripple effects of such support extend beyond immediate

---

[329] Bhattacharya, Joydeep (2023). "How to build a successful web3 community." Ninjapromo at https://ninjapromo.io/how-to-build-a-successful-web3-community.

economic benefits, contributing to the evolution of a and resilient Web3 ecosystem with the potential for sustained growth and global recognition.

Chapter 68: Networking, Startup Launch, Incubators, and Collaboration Hubs (National)

Introduction:

In the growing landscape of blockchain and Web3 technologies, fostering innovation is a collective endeavor that involves creating conducive environments for startups and businesses. This chapter explores the significance of networking programs, startup launch systems, incubators, and collaboration hubs as integral components in the quest to attract and nurture blockchain and Web3 companies.

The success of any technological ecosystem hinges on its ability to nurture and support emerging ventures. Networking programs, startup launch systems, and incubators play a pivotal role in providing the necessary resources, mentorship, and collaborative platforms that empower startups to thrive.

Networking Programs: Connecting Minds and Ideas:

Networking programs serve as catalysts for collaboration and knowledge exchange within the Web3 ecosystem.[330] By connecting minds and ideas, these programs create a vibrant ecosystem of innovation. Industry-specific events, conferences, and meetups provide valuable opportunities for entrepreneurs, developers, and investors to forge meaningful connections.

Startup Launch Systems: Igniting Entrepreneurial Sparks:

Startup launch systems act as launchpads for entrepreneurial endeavors in the Web3 space. These systems often encompass comprehensive support structures, including mentorship

---

[330] Alchemy at https://docs.alchemy.com/docs/how-to-set-up-core-web3-developer-tools (Similar)

programs, funding opportunities, and access to shared resources.[331] They are designed to guide startups through the critical early stages of development, helping them transform innovative ideas into viable businesses.

The Role of Incubators: Fostering Growth and Sustainability:

Incubators, both public and private, are crucial components of a thriving Web3 ecosystem. They offer startups a supportive environment where they can grow, iterate, and refine their products or services.[332] Incubators provide access to mentorship, co-working spaces, and, in some cases, direct funding, creating an ecosystem that fosters growth and sustainability.

Collaboration Hubs: Catalysts for Innovation:

Collaboration hubs serve as spaces where startups, established businesses, and industry experts converge to collaborate on projects and share insights.[333] These hubs, often equipped with state-of-the-art facilities and resources, provide an environment conducive to experimentation and collaboration. They are catalysts for cross-pollination of ideas and the birth of innovative solutions.

---

[331] Chainlink (2023). "The ultimate guide to launching a web3 startup." Chainlink at https://blog.chain.link/launching-a-web3-startup-guide/.

[332] Staff (2023). "A comprehensive guide toWeb3 accelerator program worldwide." Maddyness at https://www.maddyness.com/uk/2023/10/20/a-comprehensive-guide-to-web3-accelerator-programs-worldwide/.

[333] Station NYC (2023) at https://station3.nyc/.

Access to Mentorship and Expert Guidance:

One of the critical benefits offered by these support structures is access to mentorship and expert guidance.[334] Experienced mentors can provide valuable insights, share industry knowledge, and guide startups through the challenges of navigating the complex Web3 landscape. This mentorship often extends beyond technical aspects to include business development, marketing, and strategic planning.

Seed Funding and Investment Opportunities:

Financial support is a cornerstone for the growth of startups, and networking programs, startup launch systems, incubators, and collaboration hubs often facilitate access to seed funding and investment opportunities. By connecting startups with potential investors and venture capitalists, these structures contribute to the financial sustainability of innovative ventures.

Skill Development and Education:

The nature of Web3 technologies requires a skilled workforce, and these support structures contribute to skill development and education. Workshops, training programs, and educational initiatives organized within collaboration hubs and incubators equip entrepreneurs and developers with the knowledge and skills needed to navigate the complexities of blockchain and Web3.

The cumulative effect of networking programs, startup launch systems, incubators, and collaboration hubs is the establishment of a resilient and interconnected Web3 ecosystem. As startups flourish, they contribute to the ecosystem's vibrancy, attracting further talent, investment, and attention.

---

[334] Staff (2023). "Find a web3 mentor." Mentor Cruise at https://mentorcruise.com/filter/web3/.

Conclusion:

In conclusion, the success of regions in attracting and nurturing blockchain and Web3 companies is linked to the effectiveness of their networking programs, startup launch systems, incubators, and collaboration hubs. These structures create a fertile ground for innovation, providing startups with the essential elements needed for growth: mentorship, financial support, collaborative spaces, and access to a skilled talent pool. The symbiotic relationship between these support structures and the startups they incubate contributes to the evolution of a robust and Web3 ecosystem that can withstand the challenges and capitalize on the opportunities presented by the rapidly evolving technological landscape.

Chapter 69: Open Data and Accessibility (National)

Introduction:

In the modern age of Web3 technologies, nations aspiring to become hubs for innovation must recognize the pivotal role of open data and accessibility. This chapter delves into the importance of these cornerstones in attracting Web3 businesses to a nation, emphasizing their role in fostering transparency, innovation, and collaboration.

Open data refers to the idea that certain data should be freely available for use, modification, and sharing by anyone. In the context of attracting Web3 businesses, open data acts as a catalyst for innovation. By making relevant datasets accessible, governments empower businesses to derive insights, build applications, and create solutions that can drive economic growth.

Accessibility is about breaking down silos and ensuring that information is readily available to those who need it. In the Web3 landscape, where collaboration and interoperability are key, accessibility becomes a bulwark for success. When data is easily accessible, businesses can make informed decisions, researchers can conduct meaningful studies, and entrepreneurs can identify opportunities more effectively.

Transparency and Trust:

Open data cultivates transparency, a crucial element for building trust between the government, businesses, and the public. When information about government policies, economic indicators, and societal trends is openly available, businesses entering the Web3 space can operate with confidence, knowing that they have access to reliable and up-to-date information.[335]

Fueling Technological Advancements:

Web3 technologies thrive on data. From blockchain to artificial intelligence, these innovations demand vast amounts of data for training algorithms, validating transactions, and improving functionality. Nations that provide open access to relevant datasets position themselves as enablers of technological advancements, attracting businesses that require data-intensive environments.[336]

Encouraging Collaboration and Interoperability:

Accessibility promotes collaboration and interoperability by ensuring that data is not trapped within specific domains. In a Web3 ecosystem, where diverse technologies must work seamlessly together, the ability to access and share data becomes paramount. Nations that prioritize accessibility create an environment where businesses can collaborate more effectively and integrate their solutions with greater ease.

Open data and accessibility empower local innovation by democratizing information. Small and medium-sized enterprises (SMEs) and startups often lack the resources to collect extensive datasets. When governments provide open access to relevant data, these smaller entities can leverage the information to

---

[335] Ramos, Myrtle Anne (2023). "Navigating decentralized governance in Web3: Balance of Transparency and Control." Coin Telegraph at https://cointelegraph.com/innovation-circle/navigating-decentralized-governance-in-web3-balance-of-transparency-and-control.

[336] Open European Data (2023). At https://data.europa.eu/en/publications/datastories/open-data-and-ai-symbiotic-relationship-progress.

create innovative products and services, contributing to the overall vibrancy of the Web3 ecosystem.[337]

Nations that champion open data and accessibility become magnets for global talent. Entrepreneurs, developers, and professionals in the Web3 space seek environments where they can access the information necessary for their projects. By fostering a culture of openness, nations can attract a diverse pool of talent, enhancing the competitiveness of their Web3 ecosystem on the global stage.

While the benefits of open data and accessibility are substantial, it's essential to address challenges such as data privacy, security, and standardization. Governments must implement robust frameworks to protect sensitive information, ensuring that openness does not compromise individual privacy or national security.

Conclusion:

The importance of open data and accessibility cannot be overstated in the context of attracting Web3 businesses to a nation. These principles foster innovation, transparency, and collaboration, laying the foundation for a thriving Web3 ecosystem. As nations recognize the strategic significance of open data and accessibility, they position themselves as players in the global landscape of technological innovation. By embracing these principles, governments can create environments where businesses, both domestic and international, can flourish and contribute to the continuous evolution of the Web3 paradigm.

[337] Gauthier, Theo (2023). "The art of the possible with web3 interoperability" Venture Beat https://venturebeat.com/programming-development/the-art-of-the-possible-with-web3-interoperability/.

Chapter 70: Financial Inclusion (Global, Regional, National)

Introduction:

In the landscape of Web3 technologies, financial inclusion stands out as a bulwark for fostering growth and attracting businesses to a nation. This chapter explores the critical role of financial inclusion in the Web3 era, emphasizing its impact on economic development, innovation, and the creation of an inclusive digital economy.

Financial inclusion refers to the accessibility and usage of formal financial services by all segments of the population. It is not merely about providing access to traditional banking; it extends to embracing innovative financial technologies that empower individuals and businesses. In the Web3 era, where decentralized finance (DeFi) and cryptocurrencies are gaining prominence, financial inclusion becomes a cornerstone for sustainable growth.

Empowering the Unbanked and Underbanked:

A significant portion of the global population remains unbanked or underbanked, lacking access to essential financial services.[338] Web3 technologies offer an unprecedented opportunity to bridge this gap. By leveraging decentralized and blockchain-based financial solutions, nations can empower the unbanked and underbanked, providing them with tools for savings, investments, and secure transactions.[339]

---

[338] Staff (2023). "The Unbanked in America: Understanding the Challenges and Seeking Financial Inclusion." FinTwist at https://fintwistsolutions.com/blog/the-unbanked-in-america-understanding-the-challenges-and-seeking-financial-inclusion/.

[339] Contributors (2023). "Empowering the underbanked: Fostering Inclusive Financial Solutions." Finance Magnates at

Financially inclusive ecosystems stimulate entrepreneurship and innovation. When individuals and small businesses have access to financial services, they can more readily participate in economic activities. Web3 technologies, such as blockchain-based crowdfunding and decentralized lending platforms, offer alternative avenues for funding, fostering a more vibrant entrepreneurial landscape.

Decentralized Finance (DeFi) and Inclusive Access:

DeFi, a prominent component of the Web3 revolution, operates on principles of openness and accessibility. Decentralized financial platforms enable users to access a range of services, including lending, borrowing, and trading, without traditional intermediaries. By promoting DeFi solutions, nations can democratize finance, allowing citizens from all walks of life to participate in the global economy.

Financial inclusion plays a pivotal role in reducing economic disparities within a nation.[340] As Web3 technologies break down barriers to entry in the financial sector, more individuals can actively engage in economic activities. This inclusivity contributes to a more equitable distribution of resources and opportunities, fostering social cohesion and sustainable development.

Financial inclusion aligns with the adoption of cryptocurrencies and digital assets, key components of the Web3 paradigm. Cryptocurrencies provide a decentralized and borderless means of conducting transactions, making financial services accessible to anyone with an internet connection. Nations that embrace

---

https://www.financemagnates.com/fintech/payments/empowering-the-underbanked-fostering-inclusive-financial-solutions/.

[340] Staff (2023). "Is DeFi the Key to a More Equitable and Inclusive Financial Future." Zeebu at https://www.zeebu.com/blog/is-defi-the-key-to-a-more-equitable-and-inclusive-financial-future.

the use of digital currencies signal to the global community that they are at the forefront of the Web3 financial revolution.

Challenges and Regulatory Considerations:

While the benefits of financial inclusion in the Web3 era are substantial, addressing challenges such as regulatory frameworks, security, and education is crucial. Governments must strike a balance between fostering innovation and protecting consumers, ensuring that the benefits of financial inclusion are realized without compromising stability.

Conclusion:

In conclusion, financial inclusion emerges as a strategic imperative for nations aspiring to attract Web3 businesses. By prioritizing inclusive financial ecosystems, governments lay the groundwork for economic empowerment, innovation, and social progress. The Web3 era offers unprecedented opportunities to redefine financial systems and bring about positive transformations on a global scale. Nations that champion financial inclusion position themselves as pioneers in the digital economy, signaling to businesses and investors that they are committed to building a future where the benefits of Web3 technologies are accessible to all.

Conclusion Part VII

The journey through Part VII has navigated the realms of fostering a flourishing Web3 ecosystem, emphasizing the crucial factors that can shape the trajectory of nations in this digital era. As we conclude this section, it becomes evident that the convergence of technology, regulation, education, and infrastructure is pivotal in determining the success of nations in attracting and nurturing Web3 businesses.

One overarching theme has been the significance of clear and adaptive regulations. Chapters 61 through 70 underscored the imperative for nations to establish robust regulatory frameworks that accommodate the evolving landscape of Web3 technologies. The call for clarity in regulations is not merely a bureaucratic necessity but a catalyst for innovation, providing the much-needed confidence for businesses to thrive.

Regulatory sandboxes emerged as spaces for experimentation and learning. Chapter 62 delved into the role of these controlled environments in developing effective regulations. The sandbox approach allows regulators and businesses to collaborate, fostering an atmosphere where innovative Web3 solutions can be tested without the fear of stifling regulations, paving the way for a harmonious coexistence between innovation and governance.

Chapter 63 illuminated the importance of collaborations between Web3 projects and regulatory bodies. Such partnerships are vital for creating an environment where businesses can thrive without compromising the security and interests of the public. By engaging in constructive dialogue, businesses and regulators can work together to strike a balance that encourages innovation while ensuring accountability.

Chapters 64 and 65 underscored the role of education as a fundamental pillar in the Web3 journey. Regional cryptocurrency education programs and global talent attraction initiatives are crucial components in building a skilled workforce and fostering a culture of understanding around these transformative technologies.

Chapters 66 to 69 shed light on the multifaceted aspects of talent acquisition, infrastructure development, and financial inclusion. These chapters highlighted that to thrive in the Web3 landscape, nations must invest in building a robust digital infrastructure, attracting talent globally, and ensuring financial inclusivity. These factors not only stimulate economic growth but also foster an environment conducive to the flourishing of Web3 innovations.

Chapter 68 emphasized the importance of networking programs, startup launch systems, and collaboration hubs. These support systems play a crucial role in nurturing a community of innovators, providing them with the necessary resources and connections to thrive. In a world where collaboration is key, these networks become the lifeblood of a vibrant and interconnected Web3 ecosystem.

Chapters 69 and 70 explored challenges and opportunities in the form of resistance to change and the need for scalability. The business community's resistance and scalability concerns pose challenges, but they also present opportunities for nations to position themselves as leaders in overcoming these obstacles, creating an environment where Web3 technologies can truly flourish.

As we step away from Part VII, it is evident that the journey toward a mature and thriving Web3 ecosystem is complex. Nations embracing the principles of clarity, collaboration, education, and infrastructure development are poised to emerge as leaders in this digital revolution. The road ahead requires continuous adaptation, fostering an environment where innovation is not just welcomed but celebrated, and where the benefits of Web3 technologies are accessible to all.

In the chapters that follow, the exploration of the Web3 wilderness continues, delving deeper into advanced features, emerging technologies, and the transformative potential that lies ahead. The interconnected nature of these chapters reflects the ecosystem of the Web3 landscape, where each element plays a unique role in shaping the future of global economies and businesses.

Part VIII: Infrastructure Development and Cryptocurrency Exploration

Introduction Part VIII

Welcome to Part VIII of our exploration—where the digital horizon unfolds, and businesses confront the transformative challenges and opportunities of Web3. In the preceding sections, we journeyed through the realms of blockchain, decentralized finance, and the interplay between technology and commerce. Now, in this pivotal part, we delve deeper into the organizational strategies and practical implementations that define success in the Web3 era.

Web3 is not merely a technological evolution; it's a mindset shift—a departure from traditional paradigms of centralized control and a leap into the decentralized, collaborative, and innovative landscape. As we embark on chapters 71 to 80, the spotlight is on adopting and nurturing this Web3 mindset. It goes beyond technology adoption; it's about fostering a culture that thrives on continuous learning, innovation, and adaptability. Chapter 71 sets the stage by emphasizing the strategic importance of dedicated blockchain or cryptocurrency teams within an organization. These teams become the architects of digital transformation, guiding businesses through the complex terrain of blockchain integration. The ability to navigate this space requires not just technical expertise but a holistic understanding of how blockchain aligns with organizational objectives.

Chapter 72 unfolds the narrative of in-house blockchain solutions, a concept deeply aligned with the ethos of Web3. By developing solutions within the organization, businesses gain greater control and autonomy over their technological infrastructure. This strategic approach ensures tailored solutions that seamlessly integrate with existing operations while maintaining flexibility for future adaptations. The subsequent chapters (73 and 74) underscore the imperative of breaking down internal silos. Web3 is inherently collaborative, demanding the integration of fintech and blockchain solutions across various organizational departments. This interdepartmental harmony is not just a structural change; it's a cultural shift towards cross-functional collaboration and shared objectives.

The journey continues with chapters 74 and 75, shining a light on the critical role of research, development, and training. Web3 is a space where continuous learning is not just an option—it's a necessity. Organizations that invest in cultivating expertise, understanding emerging technologies, and empowering their workforce with essential skills are better positioned for success. Chapter 78 introduces a crucial theme—the imperative of robust cybersecurity measures. As businesses leverage decentralized technologies, the significance of safeguarding digital assets and sensitive information becomes paramount. Cybersecurity is not a passive defense; it's an active strategy to build trust and resilience in the face of evolving threats.

Chapter 79 unfolds the symbiotic relationship between businesses and external fintech expertise. Collaboration becomes a cornerstone of success in Web3, where businesses can leverage external knowledge to accelerate their integration of fintech solutions, leveraging the collective expertise of the broader ecosystem.

Chapters 76 and 77 provide strategic insights into incentivizing innovation within organizations. Recognizing that innovation is a bulwark for success in Web3, businesses need to cultivate environments that encourage creativity and problem-solving. Strategies that fuel innovation become essential elements in navigating the complex and evolving landscape. Chapter 80 completes this part by exploring how companies can seek fintech solutions for industry-specific problems. It delves into the versatility of Web3 technologies, showcasing their adaptability to address unique challenges across diverse industries.

As we embark on this journey through Part VIII, it's not just about embracing technology; it's about embracing the Web3 frontier as a whole. The chapters ahead will guide businesses on this transformative expedition, providing insights, strategies, and a roadmap to navigate the uncharted territories of Web3. This is an era where adaptability and innovation are not just advantageous; they are prerequisites for success. Join us as we unravel the ecosystem of Web3 and forge a path toward organizational excellence in the digital age.

Chapter 71: Establishing Fintech Teams

Introduction

In the evolving landscape of technology, businesses embracing Web3 are faced with a transformative imperative: the establishment of a dedicated blockchain and cryptocurrency team. This chapter delves into the crucial importance of assembling a specialized team poised to navigate the intricacies of the decentralized future.

As companies venture into the realm of Web3, the establishment of a specialized team is not merely an option but a strategic necessity. Blockchain and cryptocurrencies introduce a paradigm shift in how businesses operate, necessitating a deep understanding of these technologies. A dedicated team acts as the vanguard, steering the organization through the uncharted waters of decentralization.

Technical Expertise:

At the heart of a successful Web3 adoption strategy lies technical expertise. Blockchain, with its cryptographic underpinnings and decentralized architecture, requires a nuanced understanding.[341] A dedicated team brings together professionals versed in distributed ledger technologies, smart contract development, and the intricacies of various consensus mechanisms. This expertise becomes the bedrock upon which the organization can build its Web3 initiatives.

Cryptocurrencies, as integral components of many Web3 solutions, demand a nuanced understanding of their

---

[341]Hymel, Cory (2023). "How to build tech and career skills for Web3 and Blockchain." Spiceworks at https://www.spiceworks.com/tech/it-careers-skills/guest-article/building-career-skills-for-web3-and-blockchain/.

functionalities and markets.[342] A dedicated team with expertise in cryptocurrency integration, wallet development, and tokenomics ensures that the organization can leverage these digital assets effectively. From payment solutions to decentralized finance (DeFi) applications, the team becomes adept at navigating the diverse landscape of cryptocurrencies.

Security and Compliance:

Security is paramount in the Web3 space, where decentralization must coexist with robust security measures. A dedicated team is equipped to implement best practices in blockchain security, safeguarding the organization from potential vulnerabilities. Moreover, as regulations evolve in the digital asset space, the team plays a pivotal role in ensuring compliance with emerging standards, reducing legal risks associated with Web3 adoption.[343]

Innovation and Adaptability:

Web3 is a crucible of innovation, and a dedicated team serves as the catalyst for continuous advancement. By fostering a culture of innovation, organizations can stay at the forefront of technological developments. The adaptability of a specialized team ensures that the company can pivot swiftly in response to evolving industry trends, ensuring sustained relevance in the Web3 landscape.[344]

---

[342] Expertise finder (2023). At https://expertisefinder.com/find-an-expert/cryptocurrency-experts/.

[343] Staff (2023). "How to hire a chief compliance officer for your fintech." Inreg at https://www.innreg.com/blog/how-to-hire-a-chief-compliance-officer-cco.

[344] Community (2024). "How can financial technology employees improve their adaptability." Linked in at https://www.linkedin.com/advice/3/how-can-financial-technology-employees-improve-rc28f.

Web3 adoption is not confined to the IT department; it permeates across various facets of an organization. A dedicated team acts as a bridge, fostering collaboration between technical and non-technical departments. By facilitating communication and understanding, the team ensures that the benefits of Web3 technologies are harnessed throughout the organization, from supply chain management to customer relations.

Conclusion

Beyond implementation, a dedicated Web3 team serves as educational advocates within the organization. They demystify blockchain and cryptocurrencies, ensuring that stakeholders at all levels comprehend the potential and implications of Web3. This educational role is pivotal in overcoming skepticism and fostering a unified vision for the organization's digital future.

In conclusion, the establishment of a dedicated blockchain and cryptocurrency team emerges as a bulwark in a company's journey into the Web3 landscape. Beyond technical expertise, this team becomes the driving force behind innovation, security, and organizational alignment. As businesses navigate the complexities of decentralization, a specialized team stands ready to lead the charge into the future.

Chapter 72: Developing In-House Blockchain Solutions

Introduction:

As businesses look into the transformative landscape of Web3, the importance of in-house blockchain development becomes increasingly evident. This chapter unravels the strategic advantages and operational excellence that organizations can achieve by cultivating an internal capability for blockchain solutions.

In-house blockchain development affords organizations strategic control over their technological infrastructure.[345] Rather than relying solely on external solutions, businesses can tailor blockchain applications to meet their unique needs. This level of customization ensures that the technology aligns seamlessly with existing processes and can be adapted swiftly to evolving business requirements.

One of the primary benefits of in-house blockchain development lies in enhancing operational efficiency. Blockchain, as a decentralized ledger, introduces transparency and traceability into operations. By developing solutions in-house, organizations can streamline processes such as supply chain management, reducing friction and fostering a more efficient workflow.

While the initial investment in building an in-house blockchain team may seem substantial, it often proves to be a cost-effective approach in the long run. External blockchain solutions may incur recurring fees and licensing costs, which can accumulate

---

[345] Stathum, Jason (2022). "In-house v. Outsource Blockchain Development: Which do you prefer?" ReadWrite at https://readwrite.com/in-house-vs-outsource-blockchain-development-which-to-prefer/.

over time. By investing in internal capabilities, businesses gain control over their expenditures and reduce dependency on external service providers.

Security and Data Integrity:

In an era where data breaches and cyber threats are prevalent, in-house blockchain development becomes a bulwark for security and data integrity. Internal teams can implement robust security measures tailored to the organization's specific needs. The immutability of blockchain ensures that once data is recorded, it cannot be altered, providing an additional layer of protection against unauthorized access.[346]

The nature of Web3 demands agility and constant innovation. In-house development teams are better positioned to respond swiftly to emerging trends and technological advancements. This agility allows organizations to stay ahead in the competitive landscape, experimenting with new features and functionalities without the constraints of external development cycles.

Many established businesses operate with legacy systems that have been in place for years. In-house blockchain development facilitates seamless integration with these legacy systems, ensuring a harmonious transition into the Web3 era. This interoperability is crucial for businesses aiming to modernize their operations without disrupting existing workflows.[347]

Building in-house blockchain capabilities translates to knowledge retention within the organization. As the team gains expertise in blockchain technologies, this knowledge becomes an integral part of the company's intellectual capital.[348]

---

[346] Staff (2023). "The Importance of data security in Fintech." Moneye at https://moneye.io/blog/finance/the-importance-of-data-security-in-fintech.

[347] *Id.*

[348] Stathum, *supra*.

Additionally, investing in skill development for existing employees fosters a culture of continuous learning and positions the organization as a hub for technological innovation.

Conclusion:

In conclusion, the strategic decision to develop blockchain solutions in-house represents a pivotal step toward operational excellence in the Web3 landscape. Beyond cost considerations, this approach empowers organizations with strategic control, operational efficiency, heightened security, and the agility needed to thrive in an era defined by decentralization and innovation. As businesses navigate the complexities of Web3, the ability to leverage in-house blockchain development emerges as a transformative capability, positioning organizations for sustained success in the digital frontier.

Chapter 73: Integrating Fintech Across Departments

Introduction:

In the evolving landscape of Web3, the integration of fintech and blockchain technologies is not a solitary endeavor limited to a specific department; instead, it emerges as a strategic imperative that necessitates seamless integration across various facets of an organization. This chapter explores the importance of breaking down silos and fostering interdepartmental collaboration to harness the full potential of fintech and blockchain solutions.

Enhancing Operational Coherence:

Traditionally, departments within an organization operate within distinct silos, each with its own set of tools, processes, and data. The integration of fintech and blockchain technologies across departments creates a unified ecosystem where data flows seamlessly, fostering operational coherence. This unified approach eliminates redundancies, streamlines processes, and ensures a synchronized workflow across the organization.[349]

The synergy between fintech and blockchain allows for real-time data accessibility across departments. Blockchain's decentralized ledger ensures that data is updated in real-time and accessible to relevant stakeholders. This transparency facilitates better decision-making, as departments can base their strategies on the most current and accurate information available.

---

[349] Jain, Anubhav (2022). "How Fintech Startups are re-writing the future of collections." India Times at https://timesofindia.indiatimes.com/blogs/voices/how-fintech-startups-are-re-writing-the-future-of-collections/?source=app&frmapp=yes.

For businesses with customer-facing departments, the integration of fintech and blockchain contributes to an enhanced customer experience. Customer data, transaction history, and preferences stored on a blockchain enable a comprehensive understanding of customer needs.[350] Fintech tools can leverage this data to provide personalized services, ultimately improving customer satisfaction and loyalty.

The integration of fintech and blockchain across finance and accounting departments revolutionizes financial management. Smart contracts powered by blockchain can automate financial processes, reducing the risk of errors and fraud.[351] Fintech applications further enhance financial visibility, offering real-time analytics and insights that aid in strategic financial decision-making.

For organizations involved in manufacturing and distribution, integrating fintech and blockchain across supply chain and logistics departments is paramount. Blockchain ensures transparency and traceability, mitigating the risk of counterfeit products and facilitating efficient supply chain management. Fintech applications, integrated into this process, optimize inventory management and payment processes.

Cross-functional Collaboration:

Breaking down silos encourages cross-functional collaboration. The integration of fintech and blockchain necessitates collaboration between IT, finance, marketing, and other departments. This collaborative approach fosters innovation, as diverse perspectives contribute to the development and

---

[350] Mehrotra, et al (2020). A Paradigm Shift for Fintech: Why a Coherant Data Strategy has become a business necessity." Next Billion at https://nextbillion.net/fintechs-data-strategy-business/.

[351] Jain, *supra*.

implementation of comprehensive solutions that cater to the organization's overall objectives.[352]

Interdepartmental integration of fintech and blockchain aids in compliance and risk management. Blockchain's immutability ensures that compliance data is secure and tamper-proof. Fintech tools, integrated with compliance processes, automate regulatory reporting and ensure that the organization adheres to industry standards and legal requirements.[353]

As the business landscape continues to evolve, organizations must be agile and adaptive. Integrating fintech and blockchain across departments future-proofs the organization by creating a tech-savvy and responsive framework. This adaptability positions the organization to harness emerging technologies, stay ahead of industry trends, and remain competitive in the business environment.

Conclusion:

In conclusion, the interdepartmental integration of fintech and blockchain technologies represents a strategic move toward organizational synergy and efficiency. By breaking down silos and fostering collaboration, businesses can harness the combined power of these technologies to streamline operations, enhance customer experiences, optimize financial management, and ensure compliance. As organizations navigate the Web3 landscape, the seamless integration of fintech and blockchain across departments emerges as a bulwark for success, paving the way for a digitally transformed and future-ready enterprise.

---

[352] Benadiva, Josh (2023). "Bridging the gap: A Collaborative Approach to Blockchain and crypto Partnerships with Shayaan Khanna at Mastercard." Wharton Fintech at https://medium.com/wharton-fintech/bridging-the-gap-a-collaborative-approach-to-blockchain-crypto-partnerships-with-shayaan-khanna-26c2b04d030b
[353] *Id.*

Chapter 74: Creating a Fintech-Focused Research and Development Division

Introduction:

In the era of Web3, businesses face the imperative of not just adapting to change but actively leading the charge. Establishing fintech-centric research and development (R&D) teams within an organization becomes a strategic necessity. This chapter delves into the process of creating dedicated R&D teams focused on fintech, laying the groundwork for a seamless transition into the evolving Web3 landscape.

Creating fintech-centric R&D teams begins with building a foundation rooted in the understanding that innovation is a continuous journey. This foundation involves identifying key stakeholders, including experts in fintech, blockchain, and related domains.[354] These individuals form the nucleus of the R&D team, bringing diverse perspectives and specialized knowledge crucial for navigating the intricacies of Web3 technologies.

Strategic Alignment:

Strategic alignment is paramount when establishing fintech R&D teams. The objectives of these teams must align with the broader organizational goals and the envisioned transformation in the Web3 era.[355] This alignment ensures that R&D efforts are directed towards innovations that contribute meaningfully to the organization's overall strategy and digital evolution.

---

[354] Staff (2023). "R&D Tax Credits for the Fintech Industry." NeoTax at https://www.neo.tax/blog/r-d-tax-credits-for-the-fintech-industry.

[355] Staff (2024). "Strategies for Cryptocurrency Success: Bitcoin's Wealth Paradigm." Holy City Sinner at http://holycitysinner.com/strategies-for-cryptocurrency-success-bitcoin-s-wealth-parad/.

The success of fintech R&D teams hinges on the caliber of talent they attract. Organizations must actively seek individuals with a strong foundation in fintech, blockchain, data science, and other relevant fields. Additionally, fostering a culture of continuous learning is essential. Investing in the skill development of team members ensures they remain at the forefront of emerging technologies, enabling the organization to stay ahead in the fast-paced Web3 landscape.

To maximize the impact of fintech R&D, collaboration with existing departments is crucial. Establishing cross-functional collaboration frameworks enables the R&D teams to understand the pain points of different business units. By actively involving individuals from diverse departments, the R&D teams gain valuable insights that can inform the development of innovative solutions tailored to the specific needs of the organization.

Agile methodologies serve as the cornerstone of fintech R&D. The iterative and collaborative nature of agile approaches allows teams to respond swiftly to evolving requirements.[356] This agility is particularly vital in the Web3 space, where technological advancements occur rapidly. Regular sprints, feedback loops, and adaptive planning ensure that R&D efforts remain aligned with organizational objectives and markets.[357]

Infrastructure and Technology Stack:

Equipping fintech R&D teams with the right infrastructure and technology stack is imperative. This includes access to cutting-edge development tools, cloud computing resources, and blockchain platforms.[358] Providing a conducive environment

---

[356] Staff (2023). "What is Crypto Agility?" AppViewX at https://www.appviewx.com/education-center/what-is-crypto-agility/.
[357] *Id.*
[358] Staff (2023). "The Ultimate Blockchain Tech Stack Guide." Moralis at https://moralis.io/the-ultimate-blockchain-tech-stack-guide/.

empowers the teams to experiment, prototype, and implement innovative solutions seamlessly.

Prototyping and testing are integral stages in the R&D process. Fintech R&D teams should be adept at rapidly prototyping solutions and conducting rigorous testing.[359] This iterative process allows them to refine ideas, identify potential challenges early on, and ensure that the final products meet the organization's standards for functionality, security, and user experience.

The efforts of fintech R&D teams should not exist in isolation. Integration with the broader business strategy is paramount.[360] Regular communication channels between R&D teams and leadership ensure that innovations align with market trends, customer needs, and the organization's strategic direction.

Conclusion:

Establishing fintech-centric R&D teams signifies a proactive stance in preparing for the Web3 transition. By building a foundation rooted in strategic alignment, talent acquisition, and cross-functional collaboration, organizations can leverage the innovative prowess of these teams to navigate the complexities of the evolving fintech landscape. Embracing agile methodologies, investing in infrastructure, and ensuring integration with the overall business strategy positions organizations to not only adapt but thrive in the Web3 ecosystem. As the digital landscape evolves, fintech R&D teams emerge as architects of transformation, driving innovation that propels organizations into the forefront of the Web3 era.

---

[359] Anwar, Hasib (2018). "Web 3.0 Blockchain Technology Stack: The Comprehensive Guide." 101 Blockchains at https://101blockchains.com/web-3-0-blockchain-technology-stack/.

[360] Staff (2023). "The Most common and best tech stack for blockchain app development." Caper at https://gaper.io/tech-stacks-blockchain-app-development/.

Chapter 75: Employee Training Programs (Avoiding Human Error)

Introduction:

In the frontier of Web3, where security is paramount, organizations must prioritize comprehensive employee training programs. This chapter explores the critical role of such programs in fortifying teams against potential security threats and minimizing human errors. By delving into the importance of training initiatives, organizations can enhance their cybersecurity posture and navigate the complexities of the evolving digital era.

As technology evolves, the significance of the human factor in cybersecurity becomes more pronounced. Research consistently indicates that a substantial number of security breaches result from human errors, ranging from inadvertent data disclosures to falling victim to phishing attacks.[361] Acknowledging this reality is the first step toward designing effective training programs that address the unique challenges posed by human behavior in the Web3 environment.

Web3 security training should be comprehensive, covering a spectrum of topics relevant to the organization's operations. This includes understanding blockchain technology, recognizing potential threats specific to Web3, and adopting secure practices when dealing with cryptocurrencies and decentralized systems.[362] Employees should gain insights into the intricacies of

---

[361] Staff (2023). "Automating Success Reducing Human Error for Web3 Teams on Blockchain." Entendre Finance at https://entendre.finance/automating-success-reducing-human-error-for-web3-teams-on-blockchain.

[362] Lurence, Kate (2023). "Securitng Web3: Four Insights from MetaEra Summit." Linked In at https://www.linkedin.com/pulse/securing-web3-

smart contracts, decentralized finance (DeFi), and the broader blockchain ecosystem.[363]

Smart Contract & Security Awareness:

Given the prevalence of smart contracts in Web3 applications, understanding their security implications is crucial.[364] Training initiatives should provide insights into the potential vulnerabilities associated with smart contracts and how employees can contribute to securing these automated agreements. Awareness of coding best practices and auditing processes can mitigate the risks associated with smart contract vulnerabilities.

> *Phishing Attack: an attempt to steal sensitive information, generally by tricking an authorized user into giving codes or passwords for access.*

Phishing attacks remain a pervasive threat, exploiting human vulnerabilities to gain unauthorized access. Training programs should educate employees on recognizing phishing attempts, emphasizing the importance of verifying the authenticity of communication channels and refraining from clicking on suspicious links.[365] Physical-world

---

four-insights-from-metaera-summit-kate/?utm_source=rss&utm_campaign=articles_sitemaps.

[363] Staff (2023). "More than $22M Stolen from web3 platforms in October." Zokyo at https://medium.com/@zokyo.io/more-than-22m-stolen-from-web3-platforms-in-october-f33271e7fb91.

[364] Groopman, Jessica (2023). "Web 3.0 Security Risks: What you need to know." Tech Target at https://www.techtarget.com/searchsecurity/tip/Top-3-Web3-security-and-business-risks.

[365] *Id.*

simulations and practical exercises can fortify these skills, making employees more resilient against phishing attacks.[366]

The decentralized nature of Web3 does not diminish the importance of data privacy. On the contrary, it heightens the need for employees to be vigilant in safeguarding sensitive information.[367] Training should instill an understanding of data protection regulations, the secure handling of personal information, and the use of encryption technologies to maintain confidentiality in a decentralized environment.

Understanding the principles and applications of encryption is fundamental to Web3 security. Training programs should guide employees on the effective use of encryption tools, emphasizing end-to-end encryption for communications and transactions. Additionally, decentralized identity management should be a focal point, ensuring that employees comprehend the nuances of secure identity practices in a Web3 context.

Incident Response Protocols:

No security strategy is foolproof, making incident response training essential. Employees should be well-versed in incident reporting procedures, promptly recognizing and reporting any suspicious activity.[368] Simulated exercises can simulate real-world scenarios, allowing teams to practice coordinated

---

[366] Staff (2023A). "Human-Proof Finance: How Automated Accounting REsuces Web3 Teams from Errors." Entendre Finance at https://entendre.finance/human-proof-finance-how-automated-accounting-rescues-web3-teams-from-errors.

[367] Haritonova, Anastasiiya (2023). "(In)Famous Smart Contract Hacks: Why they Happen and How to Avoid Them." PixelPlex at https://pixelplex.io/blog/smart-contract-hacks/.

[368] Hammer, Yoshi (2023). "Web3 Incident response: Comprehensive Guide to Cybersecurity in Decentralized Protocols." Cyvers at https://cyvers.ai/blog/web3-incident-response-comprehensive-guide-to-cybersecurity-in-decentralized-protocols.

responses to security incidents and minimize the potential impact on the organization.

Web3 security is an evolving landscape, requiring continuous learning and adaptation. Employee training programs should not be one-time endeavors but ongoing initiatives that keep teams abreast of emerging threats and evolving best practices.[369] Regular updates and refresher courses ensure that employees maintain a high level of vigilance in the ever-changing cybersecurity landscape.

Conclusion:

Employee training programs on security and the nuances of Web3 are pivotal in fortifying organizations against the multifaceted challenges of the digital era. By recognizing the human factor in security incidents, organizations can design comprehensive training initiatives that empower teams to navigate the complexities of Web3 securely. From phishing awareness to smart contract security and incident response protocols, a well-rounded training approach positions employees as active contributors to organizational resilience in the Web3 landscape. As organizations invest in the knowledge and skills of their workforce, they not only enhance their cybersecurity posture but also foster a culture of security consciousness that permeates every aspect of their operations in the decentralized digital frontier.

[369] *Id.*

Chapter 76: Internal Adoption Strategies

Introduction:

As businesses venture into the Web3 frontier, adopting fintech solutions becomes a strategic imperative. This chapter explores effective strategies for businesses aiming to integrate Web3 technologies into their operations. From understanding the ecosystem to implementation considerations, adopting a thoughtful approach is paramount in navigating the complexities of the decentralized digital landscape.

Understanding the Web3 Ecosystem:

Before embarking on the adoption journey, businesses must comprehensively understand the Web3 ecosystem.[370] This includes gaining insights into blockchain technology, decentralized finance (DeFi), smart contracts, and other foundational elements. A nuanced understanding allows businesses to align their goals with the capabilities and opportunities presented by Web3 technologies.

A successful transition to Web3 requires a well-informed team and leadership. Educating stakeholders, from executives to operational staff, about the benefits and challenges of Web3 technologies is crucial. Workshops, training sessions, and ongoing communication channels should be established to ensure that everyone is on the same page regarding the strategic shift toward fintech in the decentralized space.[371]

Forming a dedicated cross-functional team is instrumental in navigating the complexities of Web3 adoption. This team should

---

[370] Weston, Georgia (2023). "How to build a decentralized ecosystem." 101 Blockchains. At https://101blockchains.com/build-decentralized-web3-ecosystem/.

[371] Staff (2023). "Web3 Ecosystem Explained: Guide to Tools and Platforms." Daisie at https://blog.daisie.com/web3-ecosystem-explained-guide-to-tools-platforms/.

comprise individuals with diverse expertise, including blockchain developers, legal professionals familiar with decentralized regulations, cybersecurity specialists, and individuals well-versed in traditional finance.[372] The collaboration of these experts ensures a holistic approach to Web3 integration.

Piloting Small-Scale Projects:

Instead of diving headfirst into a comprehensive Web3 overhaul, businesses can strategically start with small-scale pilot projects. These projects serve as learning experiences, allowing teams to understand the nuances of Web3 implementation, identify challenges, and iteratively improve processes. A phased approach mitigates risks and provides tangible insights for scaling up.[373]

Collaborating with established Web3 players and fintech experts can accelerate the adoption process. Partnerships with blockchain development firms, DeFi platforms, or established players in the decentralized space provide access to expertise and resources.[374] Such collaborations offer valuable insights, reduce the learning curve, and contribute to the overall success of the adoption strategy.

Web3 adoption does not necessarily mean abandoning traditional systems outright. Instead, businesses should focus on integrating Web3 technologies seamlessly with existing

---

[372] Sinha, Aashish (2023). "Web 3.0 and its ecosystem." Linked in at https://www.linkedin.com/pulse/web-30-its-ecosystem-aashish-sinha/.

[373] Moreira, Susanna (2020). "6 Small Scale Projects with large social impact." Arch Daily at https://www.archdaily.com/948459/6-small-scale-projects-with-large-social-impact.

[374] Staff (2023). "How to Build Partnerships in Web3." One Crypto Ventures at https://www.linkedin.com/pulse/how-build-partnerships-web3-one-crypto-ventures-fighe/.

infrastructure.[375] This hybrid approach allows for a smoother transition, ensuring that critical operational processes are not disrupted while gradually incorporating the benefits of decentralized technologies.

Navigating the regulatory landscape is a crucial aspect of Web3 adoption. Businesses must proactively address compliance requirements and regulatory considerations associated with decentralized technologies. Legal experts specializing in blockchain and fintech regulations should be engaged to ensure that the adoption strategy aligns with the prevailing legal frameworks.

User Education and Adoption:

Whether the target audience is internal or external users, education is paramount. Businesses should invest in user-friendly interfaces, clear documentation, and user education programs to facilitate seamless adoption.[376] Clear communication about the advantages of Web3 solutions and how users can leverage them enhances overall adoption rates.

Web3 adoption is not a one-time event; it's an ongoing process that requires continuous evaluation and iteration. Businesses should establish mechanisms for gathering feedback, monitoring performance metrics, and iterating on the adopted solutions. This iterative approach ensures that the adoption strategy remains aligned with evolving business needs and industry s.

Conclusion:

As businesses embark on the Web3 journey, adopting fintech solutions becomes a strategic imperative. By understanding the

---

[375] EDU3LABS, *supra*.

[376] Staff (2022). "Web3 in Education: Pioneering a New Era of Learning and Technology." CryptoAltruism at https://www.cryptoaltruism.org/blog/web3-in-education-pioneering-a-new-era-of-learning-and-technology.

ecosystem, educating stakeholders, building cross-functional teams, piloting small-scale projects, collaborating with established players, integrating with traditional systems, addressing compliance considerations, focusing on user education, and adopting a continuous evaluation approach, businesses can navigate the Web3 frontier successfully. These strategies collectively contribute to a holistic and effective Web3 adoption strategy, positioning businesses to thrive in the decentralized digital landscape.

Chapter 77: Incentivizing Innovation

Introduction:

In the rapidly evolving landscape of business and technology, fostering innovation is a key driver for staying ahead. This chapter explores strategies for companies to incentivize their teams towards embracing blockchain technology. From creating a culture of innovation to implementing tangible rewards, the goal is to inspire and empower teams to explore the potential of blockchain solutions.

Creating a Culture of Innovation:

Before delving into specific incentives, it's crucial to establish a culture that encourages innovation.[377] Companies should promote an environment where employees feel empowered to suggest new ideas, experiment with emerging technologies, and challenge the status quo. A culture that values creativity and risk-taking is the foundation for successful blockchain integration.

Acknowledging innovative efforts is a powerful motivator. Companies can implement recognition programs that highlight and celebrate team members who contribute groundbreaking ideas or successfully implement blockchain solutions.[378] Public acknowledgment, whether through internal communications or company-wide meetings, reinforces the importance of innovation within the organizational ethos.[379]

---

[377] Franklin, Dan (2023). "How to create a culture of Innovation in your Web3 Company." Linked In at https://www.linkedin.com/pulse/how-create-culture-innovation-your-web3-company-dan-franklin/.

[378] Community (2023). "How you can develop a web3 mindset and culture in your organization." Linked in at https://www.linkedin.com/advice/0/how-can-you-develop-web3-mindset-culture-your-organization.

[379] *Id.*

Providing professional development opportunities related to blockchain is an effective incentive. Offering training programs, workshops, or sponsoring certifications in blockchain technology allows team members to enhance their skills and become valuable contributors to blockchain initiatives.[380] This not only benefits the individuals but also strengthens the company's overall capabilities.

Incentive-Based Challenges and Competitions:

Companies can organize incentive-based challenges or competitions focused on blockchain innovation. These initiatives could involve developing blockchain-based solutions to specific business challenges or improving existing processes.[381] Offering attractive rewards, such as cash prizes, career advancements, or exclusive opportunities, stimulates healthy competition and drives innovation.

Financial Rewards and Bonuses:

Financial incentives are a classic yet potent means of motivating teams. Companies can establish bonus structures tied to successful blockchain implementations or achieving predefined milestones.[382] This direct correlation between financial rewards and successful blockchain integration serves as a tangible

---

380 Chauhan, Mudita (2023). "Fostering a culture of curiosity and innovation: They keys to succeed in Web3 era." People Matters at https://www.peoplematters.in/article/technology/fostering-a-culture-of-curiosity-and-innovation-embedding-it-in-company-values-ways-of-working-38377.

381 Oak, Eliza (2023). "Designing a Reward System for Web3 Governance." A16Crypto at https://a16zcrypto.com/posts/article/designing-reward-systems-for-web3-governance/.

382 Demirel, Omer (2023). "Redefining Token Rewards for Sustainable Web3 Economics." Medium at https://medium.com/@demirelo/redefining-token-rewards-for-sustainable-web3-economies-56e932030209.

acknowledgment of the value placed on innovative contributions.

Acknowledging that innovation often thrives in flexible environments, companies can offer flexible work arrangements as an incentive. Allowing team members the freedom to explore blockchain ideas outside traditional work structures can lead to creative breakthroughs. This approach demonstrates a commitment to fostering innovation while providing a work environment that supports individual work preferences.

In the context of blockchain, equity and token-based incentives offer a unique avenue. Companies exploring blockchain solutions can introduce token-based reward systems tied to the success of blockchain projects. These tokens could represent a stake in the project or have utility within the company's blockchain ecosystem.[383] This approach aligns the team's interests with the success of blockchain initiatives.

Dedicating specific time-off for innovation allows team members to focus solely on exploring blockchain applications without the pressures of daily responsibilities. This can be structured as periodic innovation sprints or longer-term projects, providing individuals or teams with the time and space needed for in-depth exploration and experimentation.

Cross-Departmental Collaboration:

Encouraging collaboration across departments is instrumental in fostering innovation. Companies can create cross-functional teams that bring together individuals from different disciplines to collectively explore blockchain solutions.[384] This not only

---

[383] *Id.*

[384] Staff (2023). "The benefits of Cross-Collaboration within Digital Assets." Ethos at https://medium.com/@heyethos/the-benefits-of-cross-collaboration-within-digital-asset-campaigns-f58dc004e540.

promotes diverse perspectives but also breaks down silos, fostering a collaborative culture of innovation.

Conclusion:

Incentivizing teams to embrace blockchain requires a thoughtful combination of cultural, financial, and developmental incentives. From creating a culture that values innovation to offering financial rewards, professional development opportunities, and innovative time-off structures, companies can inspire and empower their teams to explore the vast potential of blockchain technology. By recognizing and celebrating blockchain contributions, companies position themselves at the forefront of technological advancement, driving long-term success and competitiveness in the evolving business landscape.

Chapter 78: Cybersecurity Measures

Introduction:

As businesses embark on the journey into the cyberverse of Web3, the integration of cutting-edge technologies like blockchain comes with new opportunities and challenges. Among the critical considerations, cybersecurity stands out as an imperative. This chapter explores the importance of robust cybersecurity measures for companies venturing into Web3, emphasizing the protection of assets, data, and maintaining stakeholder trust.

Web3, with its decentralized and interconnected nature, introduces a paradigm shift in how businesses operate. While it offers unprecedented transparency and efficiency, it also presents unique cybersecurity challenges. Decentralization brings forth new attack vectors, and the integration of smart contracts requires specialized security measures to mitigate vulnerabilities.

Asset Protection and Data Integrity:

In the Web3 landscape, where assets are often represented in digital tokens and smart contracts govern transactions, the protection of these assets becomes paramount. Robust cybersecurity measures ensure the integrity of digital assets and prevent unauthorized access or manipulation.[385] The cryptographic principles underpinning blockchain technology contribute to securing these assets against various threats.

Smart contracts, while powerful in automating processes, require specialized security considerations. Vulnerabilities in smart contract code can lead to exploits, resulting in financial

---

[385] Staff (2024). "The role of Web3 Technology in Real-World Asset Protection." Day-By-Day at https://daybydayio.medium.com/the-role-of-web3-technology-in-real-world-asset-protection-0f8762187d14.

losses and reputational damage.[386] Companies entering Web3 must employ thorough code audits, implement best practices in smart contract development, and engage with security experts to ensure the resilience of their smart contract infrastructure.

Decentralization, a core tenet of Web3, introduces a shift in the threat landscape. Traditional security models centered around perimeter defense may prove inadequate.[387] Companies must adopt a holistic approach, focusing on endpoint security, encryption, and monitoring decentralized networks to detect and respond to potential threats effectively.

Team Trust and Reputation:

Trust is a fragile yet invaluable asset in the digital landscape. For businesses transitioning to Web3, maintaining stakeholder trust is non-negotiable. A cybersecurity breach not only poses financial risks but also erodes trust and damages the company's reputation. Proactive cybersecurity measures demonstrate a commitment to safeguarding user data and assets, bolstering stakeholder confidence.

Adherence to cybersecurity standards and regulatory requirements is fundamental. Web3 operates in a global and evolving regulatory environment. Companies must align their cybersecurity practices with applicable regulations, ensuring data protection, privacy, and compliance with industry-specific standards. This proactive approach not only mitigates legal risks but also positions the company as a responsible actor in the Web3 space.

---

[386] Ambolis, Dianna (2024). "What is real-world asset and what is the role of web3 technology in real world asset protection. "Blockchain Magazine at https://blockchainmagazine.net/real-world-asset-and-role-web3-technology/.

[387] Staff (2023). "Protection Real World Electronic Assets with Web3: Flexibility and Security." Interchainment at https://www.interchainment.com/day-by-day-web3/.

No cybersecurity strategy is complete without a robust incident response plan. Companies entering Web3 should develop comprehensive response protocols to address potential breaches promptly. From identifying and containing incidents to communicating transparently with stakeholders, a well-defined incident response plan enhances overall resilience and minimizes the impact of security incidents.

In the landscape of Web3, staying ahead of emerging threats requires continuous education and awareness.[388] Companies should invest in training programs to enhance the cybersecurity literacy of their teams. Educated employees are better equipped to recognize and respond to potential threats, contributing to the overall security posture of the organization.

Conclusion:

As businesses embrace the transformative potential of Web3, the integration of robust cybersecurity measures is not just a choice; it's a necessity. Protecting digital assets, securing smart contracts, maintaining stakeholder trust, ensuring regulatory compliance, and developing resilient incident response capabilities form the bedrock of a comprehensive cybersecurity strategy. In the evolving Web3 landscape, where innovation is coupled with responsibility, companies that prioritize cybersecurity are better positioned to navigate the challenges and reap the rewards of the decentralized future.

---

[388] *Id.*

Chapter 79: Collaboration with External Fintech Innovators

Introduction:

In the forum of Web3 integration, companies are faced with the challenge of acquiring specialized skills to navigate the complexities of blockchain technology and decentralized systems. Collaboration with external fintech companies emerges as a strategic avenue to tap into valuable skill sets. This chapter explores the benefits of collaboration in acquiring the expertise needed for successful Web3 integration.

Web3, characterized by its decentralized architecture and blockchain technology, demands a unique skill set that may not be readily available within traditional corporate structures. From blockchain development to smart contract auditing, the specialized knowledge required for seamless integration poses a challenge for companies seeking to adopt Web3.

Accessing Specialized Fintech Skills:

Collaborating with external fintech companies provides a gateway to a pool of specialized skills. Fintech firms often house experts in blockchain development, cryptography, and decentralized finance (DeFi).[389] Leveraging these external resources allows companies to access the precise skills needed for Web3 integration without the necessity of building an in-house team from scratch.

The learning curve associated with Web3 technologies can be steep, especially for companies unfamiliar with blockchain

---

[389] Staff (2023). "The Most In-Demand Skills For Fintech." Huxley at https://www.huxley.com/en-gb/blog/2020/09/the-most-in-demand-skills-for-fintech/.

principles.[390] Fintech collaborations offer an accelerated learning path. By working alongside experienced professionals, internal teams gain insights, best practices, and hands-on experience, reducing the time required to adapt to the intricacies of Web3.[391]

Every company faces unique challenges in the Web3 integration journey. Collaborating with external fintech partners enables the customization of solutions to address specific challenges.[392] Whether it is optimizing smart contract functionality or enhancing security measures, fintech collaborators bring a tailored approach, aligning their expertise with the precise needs of the partnering company.

The risks associated with Web3 integration, such as smart contract vulnerabilities or security lapses, can be mitigated through collaboration with fintech experts.[393] External partners bring a wealth of experience from diverse projects, offering insights into potential pitfalls and proactive strategies for risk management, thereby enhancing the overall security and reliability of Web3 implementations.

Cost-Efficient Web3 Adoption:

Building an in-house team of Web3 experts can be a costly and time-consuming endeavor.[394] Fintech collaborations provide a

---

[390] Harpur, Richard (2023). "5 essential fintech skills organizations need to develop now." Pluralsight at https://www.pluralsight.com/blog/it-ops/fintech-skills-to-develop.

[391] *Id.*

[392] *Id.*

[393] Staff (2022). "8 Skills That Can Help You Land a Career in Fintech." Linked In at https://www.linkedin.com/pulse/8-skills-can-help-you-land-career-fintech-spring-financial/.

[394] Staff (2023). "Web3 Application and Software Development Cost Analysis." Rising Max at https://risingmax.com/blog/cost-of-blockchain-based-app-development.

cost-efficient alternative. Companies can engage external partners on a project basis, scaling resources according to project needs. This flexibility minimizes overhead costs associated with maintaining a full-time specialized team.

Collaboration with fintech companies fosters an environment of innovation.[395] By tapping into external perspectives and experiences, companies can explore novel approaches to Web3 integration. Strategic partnerships may extend beyond immediate project requirements, creating ongoing relationships that contribute to continuous innovation and adaptation to emerging trends.

Collaborative efforts facilitate knowledge transfer from external fintech experts to the internal team. This transfer of knowledge empowers the company to build internal capacity gradually.[396] Over time, as the internal team gains proficiency, the dependency on external collaborators may decrease, ensuring a smooth transition to self-sufficiency in Web3 expertise.

Conclusion

In the ever-evolving landscape of Web3, collaboration with external fintech companies emerges as a strategic imperative for businesses aiming to integrate decentralized technologies. The benefits extend beyond accessing specialized skills to include risk mitigation, cost efficiency, and fostering a culture of innovation. By embracing fintech collaborations, companies position themselves to not only navigate the complexities of Web3 but also to thrive in the decentralized future.

---

[395] Staff (2023). "What factors affect the Cost of Web3 Development in 2023?" We Alwin at https://www.alwin.io/web3-development-cost.
[396] Staff (2024). "How Much Does Web3 Development Cost in 2024?" MaticZ at https://maticz.com/web3-development-cost.

Chapter 80: Fintech Solutions for Industry Specific Challenges

Introduction:

In the landscape of Web3, companies are increasingly recognizing the need for tailored fintech solutions to address industry-specific challenges. This chapter delves into the strategies and considerations involved in seeking and implementing fintech solutions that cater to the unique requirements of different sectors. From finance to healthcare, the exploration of industry-specific problems and the integration of targeted fintech solutions are essential for sustained business success in the Web3 era.

Identifying Industry-Specific Challenges:

Every industry faces its set of challenges, ranging from operational inefficiencies to regulatory compliance hurdles. The first step in leveraging fintech for industry-specific problems is a comprehensive analysis of these challenges.[397] This involves understanding the nuances of the industry, identifying pain points, and recognizing areas where technology can make a transformative impact.

Fintech companies specializing in industry-specific solutions often bring a wealth of knowledge and experience. Collaborating with these partners facilitates a deeper understanding of sector-specific challenges.[398] This collaboration involves a two-way exchange, where industry insights from the

---

[397] Solidity (2023). "Navigating the Web3 Technology Landscape: Challenges and Opportunities." Solidity Academy at https://medium.com/@solidity101/navigating-the-web3-technology-landscape-challenges-and-opportunities-4fd6ec37f158.

[398] Staff (2023). "Web3 Must Address Fundamental Challenges." Global Data Thematic Intelligence at https://www.verdict.co.uk/web3-fundamental-challenges/.

business are coupled with the technological expertise of the fintech partner. The result is the co-creation of solutions tailored to address specific problems.

In some cases, businesses may find generic fintech solutions that align with their industry. However, customization becomes crucial to ensure these solutions effectively address the nuances of the sector. Working closely with fintech providers to tailor generic solutions can lead to a more seamless integration that meets the specific requirements of the industry.[399]

The advent of Web3, powered by blockchain technology, introduces new dimensions to industry-specific problem-solving. Sectors such as supply chain management, healthcare, and finance can leverage the decentralized and transparent nature of blockchain to address challenges related to traceability, data security, and trust. Exploring blockchain-based fintech solutions becomes integral to unlocking novel approaches to age-old industry problems.[400]

Certain industries, like healthcare and finance, are heavily regulated. Implementing fintech solutions in these sectors requires a keen understanding of compliance requirements.[401] Fintech partners specializing in regulatory technology (RegTech) can play a pivotal role in developing solutions that not only address industry challenges but also ensure adherence to stringent regulatory frameworks.

---

[399] *Id.*
[400] Solidity, *supra*.
[401] Gavriluk, Vladislav (2023). "10 Design Challenges in Web3." Arounda at https://arounda.agency/blog/10-design-challenges-in-web3.

The Role of AI and Machine Learning in Industry Solutions:

Beyond blockchain, the integration of artificial intelligence (AI) and machine learning (ML) in fintech solutions is transformative. Industries grappling with vast datasets, such as manufacturing or logistics, can benefit from predictive analytics and automation.[402] Fintech solutions leveraging AI technologies offer predictive insights, operational efficiency, and improved decision-making tailored to industry needs.

As industries evolve, so do their challenges. Companies seeking fintech solutions for industry-specific problems must prioritize scalability and futureproofing. Solutions should be adaptable to changing markets, emerging technologies, and evolving business models. A partnership with fintech providers committed to ongoing innovation ensures that industry solutions remain relevant and effective over the long term.[403]

Conclusion:

In navigating Web3, companies stand at the crossroads of technological innovation and industry-specific challenges. The strategic integration of fintech solutions tailored to the unique demands of each sector is pivotal for sustained success. Whether through collaborative problem-solving, customization of generic solutions, or harnessing the power of blockchain and AI, businesses can unlock transformative solutions that not only address current challenges but also pave the way for future growth in the Web3 landscape.

---

[402] Staff (2023). "AI and Web3: A Symbiosis for the Future of Technology." Shardeum at https://shardeum.org/blog/ai-and-web3/.
[403] Id.

Conclusion Part VIII

In the exploration of Part VIII, we embarked on a journey through the intricacies of Web3 and its transformative impact on businesses. The chapters within this section delved into crucial aspects of adopting blockchain, fintech, and decentralized technologies to propel companies into the digital future. As we draw the curtain on this part, a comprehensive conclusion encapsulates the key takeaways and the broader implications for businesses venturing into the Web3 wilderness.

The journey through chapters 71 to 80 underscored the importance of adopting a Web3 mindset. Businesses must not only embrace technological innovations like blockchain, fintech, and decentralized finance but also foster a culture of continuous adaptation. The narrative of Web3 is not just about adopting technologies but fundamentally changing the way companies operate, innovate, and interact with their ecosystems.

Chapter 71 emphasized the pivotal role of dedicated blockchain or cryptocurrency teams within a company. As businesses navigate the complexities of Web3, having a specialized team is akin to possessing a compass in uncharted territories. These teams serve as the architects of digital transformation, ensuring that the integration of blockchain is strategic, secure, and aligned with organizational goals. Chapter 72 delved into the strategic advantages of developing in-house blockchain solutions. This approach offers businesses greater control over their technological infrastructure, enabling tailored solutions that align with specific operational requirements. The concept of in-house development resonates with the essence of Web3—decentralization and autonomy.

The significance of interdepartmental integration emerged as a recurring theme in chapters 73 and 74. Siloed approaches are antithetical to the collaborative nature of Web3. Businesses need to break down internal barriers and foster cross-functional collaboration, ensuring that fintech and blockchain solutions permeate various facets of operations, from finance to human resources. Chapters 74 and 75 highlighted the critical role of research, development, and training in Web3 adoption. Companies that invest in these areas are better equipped to understand the nuances of emerging technologies, empower their workforce with essential skills, and navigate the ever-evolving landscape of Web3.

Chapter 78 underlined the imperative of robust cybersecurity measures in the Web3 era. As businesses leverage decentralized technologies, the importance of safeguarding digital assets and sensitive information becomes paramount. Cybersecurity is not just a defensive measure but a proactive strategy to build trust and resilience in the face of evolving threats. Chapter 79 shed light on the symbiotic relationship between businesses and external fintech companies. Collaboration is a cornerstone of Web3 success, and businesses can leverage external expertise to accelerate their integration of fintech solutions and navigate the complexities of the digital age.

Chapter 76 explored strategies for incentivizing innovation, and Chapter 77 emphasized the role of incentives in driving innovation within organizations. Recognizing that innovation is a key driver of success in Web3, businesses need to cultivate environments that encourage creativity and problem-solving. In Chapter 80, we delved into how companies can seek fintech solutions for industry-specific problems. The exploration showcased the dynamism of Web3, where industries can leverage blockchain, AI, and other technologies to address unique challenges and foster innovation.

Part VIII serves as a roadmap for businesses venturing into the Web3 frontier. From blockchain teams to in-house solutions, interdepartmental integration, research, and training, the chapters outlined strategic imperatives for success. The thread that ties these insights together is a fundamental shift in mindset—an acknowledgment that Web3 is not just a technological upgrade but a paradigm shift that requires agility, collaboration, and a commitment to continuous learning. As businesses embrace the Web3 frontier, they position themselves not merely as adopters of technology but as architects of a decentralized and innovative future.

Part IX: Facing the Fears and Needs of the Community

Introduction to Part IX

Fear is a dangerous thing, and as we embark on Part IX, we need to take into consideration the problems facing the Web3 community as it relates to fear, uncertainty, and doubt (FUD). These are powerful tools which are being used to undermined the industry's reputation with potential investors and members of the community.

We start out by looking at the legitimacy of the public concern, for which there is good cause. Understanding the reason for fear allows the industry to address the reason for fear to move past the fear. One of the keyways to do this is through an aggressive education campaign. People fear what they do not known, which creates the first arm of the FUD problem.

Next, it is imperative to address the regulatory uncertainty which revolves around the space. Whether through ignorance or collusion, many nations have poorly developed financial laws- even before cryptocurrency became a concern. This means that nations and companies need to look beyond the simple integration of fintech into the community, and advance the thought toward the understanding of financial inclusion across the spectrum. This means that the industry must adapt the system to be user-friendly. This also means that the industry needs to address security risks which increase the level of uncertainty within the community.

Doubt casts the longest shadow across the face of Web3. People expect that the rich will get access to use something before they do, therefore the doubt technologies that the rich are not use. This makes adoption, public adoption, by TradFi systems imperative for rapid acceptance of the system. Government initiatives and support can enhance the speed of this adoption, giving access to more faster. This dispels the doubt through building transparency into the system.

Part XI presents that salve that is needed to calm the inflammation of FUD. Just like on any journey where you get a rash, a blister, or a cut- FUD is an annoyance that can put the entire exploration at risk. Ignoring the problem just allows it to fester and get worse- the best efforts are spent on ensuring that it is taken care of quickly. This portion of the text give businesses and regions the tools to do so in a timely manner.

Chapter 81: Understanding the Public Concerns

Introduction

In the burgeoning of Web3, where decentralization and innovation converge, businesses face not only technological challenges but the nuances of public perception. FUD—Fear, Uncertainty, and Doubt—loom as formidable adversaries that can impede the progress of Web3 initiatives.[404] This chapter delves into the importance of addressing public concerns, dispelling FUD, and shaping a positive narrative around businesses venturing into the Web3 space.

The adoption of Web3 technologies often outpaces public understanding. As businesses delve into decentralized finance, blockchain, and cryptocurrency, they encounter skepticism and apprehension.[405] This chapter recognizes the diverse array of public concerns, ranging from security and regulatory uncertainties to the perceived complexity of decentralized systems.

Transparent communication emerges as a potent antidote to FUD. Businesses need to bridge the knowledge gap by effectively communicating their Web3 initiatives, the underlying technologies, and the benefits they bring.[406] The introduction of clear, concise messaging can demystify complex concepts and build a foundation of understanding among stakeholders.

---

[404] O'Brian, Sam (2023). "How Web3 Security Concerns Might Impact You." IEEE Computer Society at https://www.computer.org/publications/tech-news/trends/web3-security-concerns.

[405] Groopman, *Supra*.

[406] O'Brian, *supra*.

Educational Initiatives as a Cornerstone:

Dispelling FUD requires proactive educational initiatives. Businesses must take the lead in educating the public, investors, and partners about the intricacies of Web3 technologies.[407] Whether through workshops, webinars, or accessible online resources, educational endeavors become pivotal in demystifying decentralized finance, blockchain, and cryptocurrency.

Security remains a paramount concern in the Web3 space. Chapter 81 underscores the significance of robust security measures and the proactive communication of these measures to the public. Demonstrating a commitment to cybersecurity builds trust and assuages fears surrounding the vulnerabilities often associated with decentralized systems.[408]

The regulatory landscape in Web3 is and, at times, uncertain. Businesses must proactively engage with regulators, advocate for clear frameworks, and communicate their adherence to compliance standards.[409] This proactive engagement serves not only to address concerns related to regulatory uncertainties but also to foster an environment of cooperation.

Conclusion

As businesses traverse the uncharted territories of Web3, the ability to manage public perception becomes a bulwark of success. Chapter 81 concludes by emphasizing that dispelling FUD is not just a defensive strategy; it's a proactive approach to building trust, fostering understanding, and establishing

---

[407] Reviredo, Tatiana (2022). "Decentralization, DAOs, and the current Web3 Concerns." CoinTelegraph at https://cointelegraph.com/news/decentralization-daos-and-the-current-web3-concerns.

[408] O'Brian, *supra.*

[409] Groopman, *supra*.

businesses as pioneers in the Web3 frontier. Through transparent communication, education, and a commitment to security and compliance, businesses can shape a narrative that aligns with the transformative potential of decentralized technologies. In the Web3 era, where collaboration and trust are paramount, addressing public concerns becomes an integral part of the journey toward widespread adoption and acceptance.

Chapter 82: Education and Awareness Campaigns

Introduction

In the modern landscape of fintech, where innovation and skepticism coexist, education emerges as the bulwark to combatting FUD—Fear, Uncertainty, and Doubt. This chapter explores the pivotal role of education as the first line of defense against misinformation, fostering a nuanced understanding of fintech concepts, and dispelling unwarranted fears.

The Complexities of Fintech

Fear, Uncertainty, Doubt: The trifecta of outside thought that interferes with business success.

Fintech, with its amalgamation of finance and technology, can be and intimidating to those unfamiliar with its intricacies.[410] Chapter 82 acknowledges the diverse set of concerns that FUD encompasses in fintech, spanning from cybersecurity threats to the perceived volatility of cryptocurrency markets.

Education serves as the cornerstone for dismantling FUD. The chapter emphasizes the need for comprehensive educational initiatives that cater to various stakeholders—ranging from the general public and investors to policymakers and businesses.[411] Establishing a foundational understanding of fintech principles is key to demystifying the complexities surrounding decentralized finance, blockchain, and emerging technologies.

---

[410] Staff (2024). "Top 8 Challenges in Fintech." Epam at https://anywhere.epam.com/business/fintech-challenges.

[411] Rootstack at https://rootstack.com/en/blog/role-fintech-education-financing-future-learning.

To combat FUD effectively, education must be accessible. Chapter 82 advocates for the creation of user-friendly learning resources, including online courses, webinars, and informative content tailored to different knowledge levels.[412] These resources aim to bridge the knowledge gap, providing individuals with the tools to navigate fintech landscapes confidently.

Businesses in the fintech space play a pivotal role in dispelling FUD. The chapter highlights the importance of proactive communication strategies employed by fintech companies. Clear and transparent messaging about their technologies, security measures, and compliance standards contributes to building trust and dispelling misconceptions.[413]

Collaboration with Educational Institutions:

Fostering collaboration between fintech companies and educational institutions is a proactive step toward dispelling FUD. When businesses engage with universities, schools, and training programs to contribute to curricula, provide industry insights, and offer real-world perspectives on fintech advancements the industry becomes more structured and more efficient as generations build on generations.[414]

Conclusion: Empowering Through Knowledge

In conclusion, Web3 must emphasize that education is the cornerstone for dispelling FUD in fintech. By investing in accessible learning resources, fostering proactive communication, and collaborating with educational institutions, businesses can empower individuals and communities to navigate the fintech landscape with confidence. As the fintech ecosystem continues to evolve, education becomes not only a

---

[412] *Id.*

[413] Epam (2024), *supra.*

[414] Corserra at https://www.coursera.org/courses?query=fintech.

tool for dispelling fears but a catalyst for embracing the transformative potential of financial technologies. In the journey toward a more inclusive, informed fintech future, education stands as the first and fundamental step.

Chapter 83: Regulatory Clarity and Consumer Protection

Introduction:

In the evolving space of fintech, regulatory clarity and robust consumer protection mechanisms are instrumental in cultivating public trust. This chapter delves into how these twin pillars not only safeguard the interests of users but also foster an environment where individuals can confidently engage with financial technologies, paving the way for broader fintech adoption.[415]

Regulatory frameworks provide the scaffolding on which fintech innovations can thrive. This Chapter explores the significance of clear and adaptive regulations, emphasizing their role in establishing a stable and secure landscape for fintech businesses.[416] It delves into the challenges posed by ambiguous regulations and how regulatory clarity acts as a catalyst for innovation while ensuring compliance.

Consumer protection is at the heart of building trust in fintech. The chapter delineates the various dimensions of consumer protection, from secure transactions and data privacy to fair lending practices. It underscores the need for fintech companies

---

415 Staff (2023). "Why fintech and crypto companies are craving good regulation." Fast Company at https://www.fastcompany.com/90902044/why-fintech-and-crypto-companies-are-craving-good-regulation.

416 Bains, Parma & Wu, Caroline (2023). "Institutional Arrangements for Fintech Regulations: Supervisory Monitoring." IMF at https://www.elibrary.imf.org/view/journals/063/2023/004/article-A001-en.xml.

to prioritize user safety and design platforms with robust security measures and transparent policies.[417]

Fostering Regulatory Innovation:

In an era of rapid technological advancements, regulatory frameworks must evolve to keep pace with innovation. Chapter 83 advocates for regulatory bodies adopting an innovative mindset, embracing technologies like blockchain and AI to enhance surveillance, compliance, and enforcement. The synergy between fintech innovation and regulatory evolution can create a harmonious and ecosystem.

While fostering innovation, it is crucial to strike a balance between promoting growth and safeguarding against potential risks. The chapter discusses the delicate equilibrium that regulators must maintain, ensuring that fintech developments are aligned with broader financial stability goals and societal interests.

A knowledgeable user is an empowered user. Chapter 83 emphasizes the role of educational initiatives in informing consumers about their rights, the functionalities of fintech services, and the protections afforded by regulations. Informed consumers contribute to a healthier fintech ecosystem by making sound financial decisions and demanding accountability from service providers.[418] Drawing on case studies, the chapter illustrates instances where regulatory clarity and robust consumer protection measures have led to positive outcomes.

---

[417] Rhodes, Gareth (2023). "Regulatory Clarity? Financial Watchdogs could not be more clear." Coin Desk at https://www.coindesk.com/consensus-magazine/2023/01/27/regulatory-clarity-financial-watchdogs-could-not-be-more-clear/.

[418] Staff (2023). "Fintech Regulation Approaches: What Tech Model Could be Used to Regulate the Fintech Space." Chargebacks 911 at https://chargebacks911.com/fintech-regulation-approaches/.

These real-world examples highlight the symbiotic relationship between effective regulations, consumer trust, and the flourishing of fintech innovations.

Conclusion

This chapter concludes by asserting that trust is the primary factor of a successful fintech ecosystem. Regulatory clarity and consumer protection, when implemented thoughtfully, foster an environment where fintech innovations can thrive while ensuring the well-being of users. As the fintech landscape continues to unfold, regulatory frameworks and consumer-centric policies stand as pillars that uphold the integrity of the industry, paving the way for widespread adoption and sustained growth.

Chapter 84: Addressing Privacy Concerns

Introduction

As businesses embrace the transformative potential of Web3 technologies, the paramount issue of privacy comes to the forefront. Chapter 84 dives into the complex landscape of privacy concerns within Web3, examining the challenges, emerging solutions, and the crucial role businesses play in ensuring a delicate balance between innovation and safeguarding user data.

The Unfolding Web3 Privacy Landscape:

Web3, with its decentralized and interconnected nature, introduces a paradigm shift in how individuals interact with digital platforms.[419] The distinctive privacy challenges arising from increased connectivity, blockchain transparency, and the proliferation of decentralized applications (dApps) present a unique situation within the regulatory and consumer world.[420] It underscores the need to address these challenges proactively to build and maintain user trust.

One of the key promises of Web3 is enhanced user anonymity through decentralization. However, this very feature poses challenges in terms of regulatory compliance and mitigating illicit activities. The chapter explores the delicate balance between providing users with privacy and ensuring responsible use of decentralized technologies.[421]

---

[419] Staff (2023). "What is Privacy on Web3?" Fishy on chain at https://medium.com/web3-insights/what-is-privacy-in-web3-3226ac75f2d1.

[420] Howell, James (2023). "The Promising Impact of Web 3.0 on Data Privacy." 101 Blockchains at https://101blockchains.com/web3-impact-in-data-privacy/.

[421] Shardeum, *supra*.

Smart contracts, a hallmark of Web3, operate on a public ledger, bringing transparency to transactions. While this transparency fosters trust, it also raises concerns about user privacy. Here we examine the nuances of smart contracts, discussing the trade-offs between transaction visibility and protecting sensitive user information.

In the Web3 landscape, data ownership and control shift towards users. The chapter delves into the empowering aspects of users having more control over their data, exploring decentralized identity solutions and user-centric approaches to data management.[422] It emphasizes how businesses can champion these initiatives to align with evolving privacy expectations.

Emerging Privacy Solutions:

In response to heightened privacy concerns, innovators within the Web3 space are developing cutting-edge privacy solutions. Zero-knowledge proofs, privacy-focused blockchain networks, and other cryptographic techniques aimed at enhancing user privacy are of paramount importance to ensuring that the field is fully functional and can gain the consumer's trust.[423] It underscores the importance of businesses adopting and contributing to these solutions.

Regulatory Compliance in a Decentralized World:

The decentralized nature of Web3 challenges traditional regulatory frameworks designed for centralized entities. This chapter discusses the evolving regulatory landscape, the role of self-regulation within the Web3 community, and the need for

---

[422] *Id.*

[423] Kuznetsov, Nikolai (2023). "Fostering Privacy the Right Way: Web3 Should Make Data Privacy a Default." Nasdaq at https://www.nasdaq.com/articles/fostering-privacy-the-right-way%3A-web3-should-make-data-privacy-a-default.

collaboration between industry stakeholders and regulators to establish coherent privacy standards.[424] Education is a cornerstone in addressing privacy concerns. The chapter emphasizes the role of businesses in educating users about the intricacies of Web3 technologies, the privacy features embedded in decentralized systems, and best practices for securing personal information in this new paradigm.

Conclusion:

This chapter concludes by highlighting that addressing privacy concerns in the Web3 space is not a one-size-fits-all endeavor. Instead, it requires a multi-faceted approach involving technological innovation, regulatory adaptability, and user education. As businesses navigate this evolving landscape, prioritizing privacy becomes not only a legal and ethical imperative but a strategic move to build enduring trust in the Web3 era. By championing privacy-centric practices, businesses can contribute to shaping a future where Web3 technologies coexist seamlessly with robust privacy protections.

---

[424] *Id.*

Chapter 85: Fintech for Financial Inclusion

Introduction

In an era dominated by digital advancements, this chapter delves into the transformative role of fintech in fostering financial inclusion. From unbanked populations to underserved communities, fintech emerges as a powerful tool to bridge gaps, providing access to financial services and empowering individuals worldwide.

Unlocking Access to the Unbanked:

Millions around the globe remain unbanked, excluded from traditional financial systems. This chapter explores how fintech solutions, leveraging mobile technologies and innovative payment platforms, are breaking down barriers. By offering accessible, low-cost alternatives, fintech is empowering the unbanked to participate in the formal economy.[425]

Digital payment systems, facilitated by fintech, are instrumental in fostering inclusive economies. The chapter examines the impact of mobile wallets, digital currencies, and peer-to-peer payment platforms in providing secure, efficient, and affordable financial transactions, particularly in regions with limited access to traditional banking infrastructure.

Microfinance and Fintech Synergy:

The synergy between microfinance and fintech has the potential to revolutionize access to credit for small entrepreneurs and individuals with limited resources. This chapter explores how fintech-driven microfinance initiatives are empowering individuals to start or expand small businesses, thereby contributing to economic growth in underserved

---

[425] Staff (2023). "Web3's Potential: Reaching Untapped Markets." Blockmate at https://www.linkedin.com/pulse/web3s-potential-reaching-untapped-markets-blockmate/.

communities.[426] Financial inclusion goes beyond banking services. The chapter discusses how insurtech, a subset of fintech, is playing a pivotal role in providing affordable insurance solutions to populations traditionally excluded from coverage. By leveraging technology, insurtech is mitigating risks and providing a safety net for vulnerable communities.

Promoting financial literacy is integral to fostering inclusion. This chapter explores how fintech is not just providing financial services but also educating users on financial management. From budgeting apps to online financial education platforms, fintech is enhancing financial literacy and empowering individuals to make informed decisions.[427] Blockchain technology, with its decentralized and transparent nature, has the potential to revolutionize inclusive finance. The chapter discusses how blockchain-powered solutions, such as decentralized finance (DeFi), are eliminating intermediaries, reducing costs, and providing access to a wide array of financial services for underserved populations.[428]

The role of governments in fostering financial inclusion is pivotal. This chapter explores how public-private partnerships, with fintech at the forefront, can amplify the impact of government initiatives. Digital identity systems, subsidy disbursements, and fintech-driven social welfare programs are

---

[426] Staff (2023). "Cracking Open Financial Freedom: AI, Web3, and Crypto for the Unbanked." Crypto with Kieron at https://medium.com/@cryptowithkieron/cracking-open-financial-freedom-ai-web3-and-crypto-for-the-unbanked-8af81981e86f.
[427] *Id.*
[428] Svensson, Conor (2023). "Unintended Victims of Fintech's Quest for Scale. How Web3 Offers a Lifeline." Web3Labs at https://blog.web3labs.com/fintech-the-unbanked-problem-and-the-emergent-crypto-solution.

discussed as instruments of inclusive governance.[429] While fintech brings immense opportunities for financial inclusion, it also poses challenges. Here we address issues of data privacy, security, and ethical considerations in the deployment of fintech solutions. Striking a balance between innovation and responsible practices is crucial for ensuring sustainable and equitable financial inclusion.[430]

Conclusion

In conclusion, this chapter underscores the pivotal role of fintech in building an inclusive financial landscape. By embracing technological innovation, fostering collaboration, and addressing ethical considerations, fintech emerges as a catalyst for positive change, ensuring that financial services reach every corner of the globe. As businesses and governments continue to leverage fintech for inclusive growth, they contribute to a future where financial empowerment becomes a reality for all.

---

[429] *Id. See Also* Faury, Jerome (2023). "Supporting the Unbanked." Immersve at https://medium.com/@Immersve/supporting-the-unbanked-b6cd1d554424.

[430] Blockmate, *supra*.

Chapter 86: User-Friendly Interfaces and Accessibility

Introduction

In the fast-evolving realm of fintech, the user interface (UI) stands as the gateway to seamless financial experiences. This chapter looks into the critical importance of a user-friendly interface for fintech websites. As users increasingly engage with financial services online, the design and functionality of these interfaces become paramount for fostering trust, engagement, and ultimately, user satisfaction.

Understanding User-Centric Design:

At the core of a user-friendly fintech interface is a design philosophy centered around user needs and expectations. This chapter explores how user-centric design principles prioritize simplicity, intuitiveness, and accessibility. By understanding the user journey and preferences, fintech platforms can create interfaces that resonate with a diverse user base.

First impressions matter, especially in the fintech space. It is essential that a business discusses the significance of streamlining onboarding processes through intuitive interfaces.[431] From account creation to identity verification, a user-friendly UI ensures that users can effortlessly navigate through the necessary steps, minimizing friction and fostering a positive onboarding experience.

Financial transactions demand a high level of transparency. This chapter explores how a user-friendly fintech interface enhances financial transparency by presenting information in a clear and

---

[431] Srivastav, Soarbh (2023). "Designing User-Centric Blockchain Wallets: Enhancing Security and Usability." Bootcamp at https://bootcamp.uxdesign.cc/designing-user-centric-blockchain-wallets-enhancing-security-and-usability-9d0da3dcb882.

comprehensible manner.[432] From transaction histories to account statements, transparent interfaces build trust and empower users to make informed financial decisions.

Navigating a fintech platform should be intuitive. The chapter discusses the importance of logical navigation structures and easily discoverable features. Through thoughtful UI design, fintech websites can guide users seamlessly through various functionalities, ensuring that users can leverage the full spectrum of financial services without confusion.

Personalization for User Engagement:

The era of one-size-fits-all interfaces is evolving into personalized, adaptive experiences. A business that explores how fintech platforms can leverage data-driven insights to tailor interfaces to individual user preferences will be more successful than one that does not. Personalization not only enhances user engagement but also contributes to a sense of ownership and relevance in the financial journey.[433]

As users engage with fintech services across a myriad of devices, ensuring a consistent and responsive design is crucial. This chapter discusses the significance of responsive web design, where interfaces adapt seamlessly to different screen sizes and device types.[434] A uniform experience across devices contributes to accessibility and user satisfaction.

---

[432] Corcoran, Will (2022). "Stacks overview: Web3 by Bitcoin." Designlab at https://designlab.com/blog/crypto-ux-design/.

[433] Garavaglia, Roberto (2023). "Web3 Marketing: revolutionizing customer engagement in the hospitality industry." Hospitalitynet at https://www.hospitalitynet.org/opinion/126002212.html.

[434] Silver, Charlie (2023). "Web3's impact on personalization, trust, and engagement through time honored-strategies." Coin Telegraph at https://cointelegraph.com/innovation-circle/web3s-impact-on-personalization-trust-and-engagement-through-time-honored-strategies.

Balancing user-friendliness with robust security measures is a delicate yet imperative task. The chapter explores how fintech interfaces can integrate advanced security features without compromising the user experience. From two-factor authentication to biometric recognition, maintaining a secure environment is integral to user trust.

User-friendly interfaces are not static; they evolve based on user feedback and technological advancements.[435] This chapter discusses the importance of feedback loops, user testing, and continuous improvement cycles. Fintech platforms that actively seek and implement user insights are better positioned to adapt to evolving user expectations.

Conclusion

In conclusion, this chapter underscores that the journey through fintech should be both empowering and effortless. A user-friendly interface is not just a design preference; it is a strategic imperative. By prioritizing simplicity, transparency, and personalization, fintech platforms can shape positive financial experiences, ensuring that users navigate the complexities of finance with confidence and ease.

---

[435] *Id.*

Chapter 87: Mitigating Security Risks

Introduction:

Web3, the decentralized and interconnected era of the internet, presents a transformative landscape for fintech and blockchain. However, with innovation comes the challenge of mitigating security risks.[436] This Chapter delves into the intricacies of securing the web3 environment, exploring strategies to safeguard user data, transactions, and the overall integrity of decentralized systems.

Securing web3 involves a departure from traditional security models. This chapter discusses how the decentralized nature of web3, facilitated by blockchain technology, introduces new challenges and opportunities.[437] From securing smart contracts to protecting user identities, the security paradigm in web3 demands a holistic and innovative approach.

Decentralized Identity Management:

Self-Sovereign Identity (SSI)- modeling which allows for the management of digital identities where businesses or individuals have sole ownership over the ability control systems.

Web3's emphasis on decentralization extends to identity management. This chapter explores decentralized identity solutions, discussing the role of self-sovereign identity (SSI) and blockchain-based

---

[436] Staff (2024). "What is Web 3.0: A Beginners Guide to the Decentralized Internet of the Future." Coin Telegraph at https://cointelegraph.com/learn/what-is-web-3-0-a-beginners-guide-to-the-decentralized-internet-of-the-future.

[437] EY at https://www.ey.com/en_us/financial-services/how-will-web3-and-the-metaverse-create-opportunities.

identity verification.[438] By empowering users with control over their identity data, web3 enhances privacy while mitigating risks associated with centralized identity repositories.

Smart contracts, integral to many web3 applications, require meticulous auditing. The chapter delves into the importance of smart contract security audits and outlines best practices for developers. From avoiding common vulnerabilities to implementing standardized security protocols, thorough auditing ensures the reliability and trustworthiness of decentralized applications (DApps).[439]

Web3 relies on consensus algorithms to validate transactions and secure the network. This chapter discusses the security considerations associated with popular consensus mechanisms like Proof-of-Work (PoW) and Proof-of-Stake (PoS). By understanding the strengths and vulnerabilities of these algorithms, stakeholders can make informed choices to enhance network security.

Cryptography is the bedrock of web3 security. The chapter explores the role of cryptographic principles in securing data, transactions, and communications. Topics include end-to-end encryption, cryptographic hashing, and the integration of zero-knowledge proofs for enhanced privacy.[440] Understanding cryptographic mechanisms is fundamental to mitigating risks in a decentralized environment.

---

[438] Gillis, Alexander (2023). "Self-Sovereign Identity." Tech Target at https://www.techtarget.com/searchsecurity/definition/self-sovereign-identity.
[439] Frankenfield (2023A), *supra*.
[440] Staff (2023). "Blockchain Consensus Mechanisms beyond PoW and PoS." Cryptopedia at https://www.gemini.com/cryptopedia/blockchain-consensus-mechanism-types-of-algorithm.

Securing Decentralized Finance (DeFi) Platforms:

Decentralized finance is a cornerstone of web3 innovation, but it introduces unique security challenges. This chapter discusses strategies to secure DeFi platforms, addressing issues such as flash loan attacks, smart contract vulnerabilities, and the importance of comprehensive risk assessments in decentralized financial ecosystems.

Web3 networks rely on a distributed network of nodes to maintain consensus. This chapter explores measures to secure these nodes, including firewall configurations, DDoS protection, and the implementation of secure communication protocols. Protecting the network infrastructure is essential for ensuring the integrity of decentralized systems.[441]

Mitigating security risks in web3 requires collaboration between developers and end-users. The chapter emphasizes the importance of user education and security awareness programs. By fostering a security-conscious user base, web3 platforms can reduce the likelihood of social engineering attacks and enhance the overall resilience of the ecosystem.

Conclusion:

In conclusion, this chapter underscores that the decentralized future of web3 demands proactive and adaptive security measures. From decentralized identity solutions to securing smart contracts and educating users, mitigating security risks is a collective effort. As web3 continues to redefine the digital landscape, safeguarding its foundations ensures a future where decentralized systems can thrive securely and resiliently.

---

[441] Smithmyer (2023A), *supra*.

Chapter 88: Collaboration with Traditional Finance.

Introduction:

As the web3 era reshapes the financial landscape, the importance of collaboration between decentralized technologies and traditional finance becomes increasingly evident. This chapter provides a comprehensive overview of the significance of forging alliances between the emerging web3 ecosystem and the established world of traditional finance.

Web3's Innovation and Traditional Finance's Stability:

Web3 technologies, powered by blockchain and decentralized principles, bring innovation and efficiency to financial processes. However, they coexist with traditional financial systems that offer stability, regulatory frameworks, and a wealth of experience. This chapter explores how collaboration can harness the strengths of both, creating a symbiotic relationship that fosters progress without compromising stability.[442]

The chapter delves into how traditional financial institutions can benefit from integrating decentralized technologies like blockchain.[443] From enhancing security through distributed ledger systems to optimizing processes with smart contracts, web3 offers tools that can address inefficiencies and streamline operations within traditional financial structures.[444]

---

[442] Akyaw, Elisha Owusu (2023). "Hashing it out: How Web3 is causing an evolution of traditional finance." Coin Telegraph at https://cointelegraph.com/news/hashing-it-out-how-web3-is-causing-an-evolution-of-traditional-finance.

[443] *Id.*

[444] Hooper, Anatol (2023). "How User-friendly Web3 wallets bridge the gap between TradFi and DeFi." Coin Telegraph at https://cointelegraph.com/news/how-user-friendly-web3-wallets-bridge-the-gap-between-tradfi-and-defi.

Web3's decentralized nature facilitates faster and more cost-effective cross-border transactions. The chapter discusses how collaboration can leverage blockchain to overcome the challenges associated with traditional international payment systems.[445] By integrating web3 solutions, financial institutions can enhance the speed, transparency, and traceability of global transactions.

Exploring Decentralized Finance (DeFi) Integration:

Decentralized finance (DeFi) has emerged as a prominent aspect of web3, offering innovative financial services without intermediaries.[446] This chapter examines how collaboration with traditional finance can lead to the integration of DeFi protocols, potentially expanding access to financial services and creating new opportunities for a broader range of users.[447]

Regulatory compliance is a cornerstone of traditional finance, ensuring consumer protection and system stability.[448] The chapter explores how collaboration with web3 can navigate regulatory challenges, emphasizing the need for dialogue between decentralized projects and regulatory authorities to create frameworks that foster innovation while maintaining compliance.

Many traditional financial institutions operate on legacy systems that may not be inherently compatible with decentralized

---

[445] Akyaw, *supra*.

[446] Casey, Michael (2023). "Decentralized and Tradiational Finance can thrive in tandem to fund renewable energy and other pressing needs, but only with clear standards and rules." IMF at https://www.imf.org/en/Publications/fandd/issues/2022/09/Point-of-View-Defi-Tradfi-must-work-together-Michael-Casey.

[447] *Id*.

[448] Staff (2023). "TradFi and DeFi: Key Differences." Hedra at https://hedera.com/learning/decentralized-finance/tradfi.

technologies.[449] This chapter discusses strategies for integrating blockchain into existing infrastructure, allowing for a smooth transition and enabling the benefits of blockchain without requiring a complete overhaul of established systems.[450]

Enhancing Financial Inclusion:

Collaboration between web3 and traditional finance can contribute to financial inclusion. By leveraging decentralized technologies, financial institutions can reach underserved populations, offering services like digital identity verification, microfinance, and peer-to-peer lending. The chapter explores how this collaboration can bridge gaps in access to financial services.

The chapter candidly addresses the challenges associated with collaboration, including the cultural shift required within traditional financial institutions. Simultaneously, it highlights the opportunities that arise when combining the innovation of web3 with the experience and stability of traditional finance, fostering an environment where both can thrive.

Conclusion:

This Chapter underscores the transformative potential that arises from collaborating with traditional finance in the web3 era. By recognizing the strengths and addressing the challenges, this collaboration paves the way for a synergistic future where decentralized technologies and traditional financial systems complement each other, creating a more inclusive, efficient, and robust global financial ecosystem.

---

[449] Casey, *supra*.
[450] *Id*.

Chapter 89: Government Initiatives and Support

Introduction:

In the field of fintech, the role of governments becomes pivotal in driving innovation, fostering economic growth, and expanding financial technology into new regions.[451] This chapter explores the significance of government initiatives and support in shaping the future of fintech on a global scale.

The chapter begins by highlighting the transformative potential of fintech in revolutionizing traditional financial services. Governments, recognizing the impact of technological advancements, increasingly view fintech as a catalyst for financial inclusion, economic development, and job creation. The introduction sets the stage for understanding how proactive government involvement can unlock these benefits.

Building Regulatory Frameworks for Innovation:

Governments play a crucial role in establishing regulatory frameworks that balance innovation with consumer protection and system stability.[452] The chapter delves into how regulatory clarity and adaptive frameworks contribute to creating an environment where fintech companies can thrive. Case studies and examples illustrate successful models adopted by governments to nurture innovation while mitigating risks.

One of the key drivers explored in this chapter is the role of government incentives and funding support. By offering financial incentives, tax breaks, and grants, governments can encourage fintech startups and established companies to invest

---

[451] Copland, Tim (2023). "Why is DeFi so Important?" The Block at https://www.theblock.co/learn/245713/why-is-defi-so-important.
[452] *Id.*

in research, development, and expansion.[453] The chapter examines successful instances where such incentives have spurred growth and positioned regions as fintech hubs.

Government initiatives can significantly impact financial inclusion through fintech. The chapter explores how policies aimed at reaching underserved populations, fostering digital literacy, and promoting inclusive financial services contribute to broader economic participation.[454] Case studies highlight successful government-led programs that have enhanced financial access for marginalized communities.

Collaboration Between Governments and Fintech Ecosystem:

Successful expansion of fintech often involves collaboration between governments and the fintech ecosystem. The chapter discusses how partnerships, regulatory sandboxes, and innovation hubs foster a collaborative environment.[455] This collaborative approach enables governments to stay informed about technological advancements, while fintech companies benefit from a supportive ecosystem.[456]

---

[453] Staff (2023). "Navigating the Regulatory Waters of Web3 Future: How Business can Harness the Power of Evolving Legal Frameworks in Blockchain." Crypto News Hub at https://www.linkedin.com/pulse/navigating-regulatory-waters-web3-future-how-businesses/?trk=article-ssr-frontend-pulse_more-articles_related-content-card.

[454] Kerai, Prakash; Nightingale, Paul & Gohil, Khillian (2023). "Innovation v. Regulation" Teck UK at https://www.techuk.org/resource/innovation-vs-regulation-web3.html.

[455] Engler, Henry (2023). "Bank-Fintech partners come under closer scrutiny from US bank regulators." Thompson Reuters at https://www.thomsonreuters.com/en-us/posts/investigation-fraud-and-risk/bank-fintech-partnerships/.

[456] *Id.*

Global Competitiveness and Economic Growth:

Government support for fintech contributes to a region's global competitiveness and economic growth.[457] By embracing digital transformation and encouraging fintech innovation, governments position their economies as and attractive for international investors.[458] The chapter analyzes how strategic government interventions can lead to increased foreign direct investment and job creation.

While government support is crucial, the chapter also addresses potential challenges and risks associated with government involvement in fintech.[459] Striking the right balance between oversight and innovation is explored, along with strategies for mitigating risks and ensuring a sustainable fintech ecosystem.

Conclusion:

In conclusion, this chapter emphasizes that government initiatives and support are indispensable for charting a fintech-forward future. By adopting proactive policies, regulatory frameworks, and investment incentives, governments can position their regions as vibrant fintech hubs, driving economic growth, fostering innovation, and ensuring that the benefits of fintech reach all corners of society. This chapter envisions a collaborative future where governments and the fintech industry work hand in hand to create a global landscape marked by inclusive and innovative financial services.

---

[457] Tang, Song; Chen, Zejin; Chen, Jiawei; Quan, Lei & Guan Kiayang (2023). "Does Fintech promote corporate competitiveness? Evidence from China." Finance Research Letters at https://www.sciencedirect.com/science/article/abs/pii/S1544612323010322.

[458] Staff (2023). "Fintechs: A New Paradigm in Growth." Mckinsey at https://www.mckinsey.com/industries/financial-services/our-insights/fintechs-a-new-paradigm-of-growth.

[459] Tang et at, *supra*.

Chapter 90: Building Trust Through Transparency

Introduction:

In the web of the cryptoverse, trust stands as a pillar for widespread adoption and sustained growth. This chapter, titled "Building Trust in the Cryptoverse through Transparency," explores the pivotal role of transparency in establishing trust within the crypto and blockchain ecosystem.

The chapter begins by acknowledging the historical challenges faced by the cryptoverse in terms of trust. From the early days of skepticism and association with illicit activities to concerns over security breaches and market manipulation, the cryptoverse has grappled with a trust deficit. This introduction sets the stage for dissecting how transparency becomes a potent antidote to these issues.

Transparency in Blockchain:

Blockchain, the underlying technology of most cryptocurrencies, is inherently transparent. The decentralized and immutable nature of blockchain ensures that transactional data is openly accessible. The chapter delves into how this transparency fosters trust by providing users with verifiable and unalterable records of transactions.[460] Examples and case studies illustrate instances where blockchain transparency has enhanced trust.[461]

---

[460] Hayes, Adam (2023). "Learn what these digital public ledgers are capable of." Investopedia at https://www.investopedia.com/terms/b/blockchain.asp.

[461] Staff (2023). How transparency through blockchain helps the cybersecurity community." IBM at https://www.ibm.com/blog/how-transparency-through-blockchain-helps-the-cybersecurity-community/.

Another facet explored is the concept of open-source development within the cryptoverse. The collaborative nature of open-source projects allows developers worldwide to scrutinize, contribute, and improve code.[462] The transparency of open-source development builds trust among users, as they can be assured that the technology is subject to continuous peer review and improvement.

Tokenomics, the economic model governing cryptocurrencies, is dissected to showcase how transparency in token distribution, circulation, and governance mechanisms contributes to trust.[463] By elucidating how tokens are created, distributed, and utilized within a blockchain ecosystem, users can make informed decisions, thereby fostering a more transparent and trustworthy environment.

Regulatory Compliance and Reporting:

The chapter explores the critical role of regulatory compliance and transparent reporting mechanisms in building trust.[464] Cryptocurrency projects and exchanges adhering to regulatory standards not only comply with legal requirements but also demonstrate a commitment to transparency. Examining instances where regulatory compliance has positively influenced trust, the chapter illustrates the symbiotic relationship between transparency and regulatory adherence.[465]

---

[462] Hayes, *supra*.

[463] Gaur, Vishal & Gaiha, Abhinav (2020). "Building a Transparent Supply Chain." Harvard Business Review at https://hbr.org/2020/05/building-a-transparent-supply-chain.

[464] Staff (2023). "The Use of Blockchain in Compliance and Regulatory Reporting." Reintech at https://reintech.io/blog/blockchain-compliance-regulatory-reporting.

[465] Deloitte at https://www2.deloitte.com/uk/en/pages/financial-services/articles/does-blockchain-mean-an-end-to-regulatory-reporting.html

Community Engagement and Communication:

Effective communication and community engagement emerge as essential components of transparency. The chapter discusses how crypto projects that actively communicate their objectives, updates, and challenges to the community foster trust. Real-world examples underscore the impact of transparent communication in maintaining user confidence during volatile market conditions or project developments.

Acknowledging that challenges exist, the chapter addresses concerns such as privacy, the balance between transparency and decentralization, and potential manipulations. By recognizing and proactively addressing these challenges, the cryptoverse can enhance overall transparency, reinforcing trust among participants.

Conclusion

In conclusion, Building Trust in the Cryptoverse through Transparency envisions a future where transparency becomes the norm rather than the exception. The chapter underscores that transparency is not merely a choice but a necessity for the sustainable growth of the cryptoverse. By embracing transparency in technology, economics, compliance, and communication, the cryptoverse can cultivate an environment where trust thrives, paving the way for mainstream adoption and integration into global financial systems.

Conclusion Part IX

In the landscape of Part IX, "Facing the Fears and Needs of the Global Community," we embarked on a journey through the intricacies and challenges of the cryptoverse. This concluding reflection encapsulates the key takeaways and the broader significance of the chapters exploring trust, transparency, and the diverse aspects of fintech.

The chapters in Part IX ventured into the heart of global concerns surrounding fintech and blockchain. From dispelling fear, uncertainty, and doubt (FUD) to addressing issues of education, regulatory clarity, and user-friendly interfaces, each chapter sought to demystify the complex world of the cryptoverse. By dissecting these concerns, we laid the groundwork for understanding the evolving s of fintech adoption on a global scale.

A recurring theme throughout Part IX has been the paramount importance of trust. Trust is not a mere luxury but a necessity for the flourishing of the cryptoverse. Whether it is building trust through transparency in blockchain technology, open-source development, or regulatory compliance, the chapters collectively underscored that trust is the bedrock upon which the future of fintech stands.

Transparency emerged as the guiding light illuminating the path toward a more inclusive and trustworthy cryptoverse. The exploration of blockchain's inherent transparency, open-source development, and economic models laid the foundation for building confidence among users and stakeholders. Transparency is not just an ideal but a pragmatic solution to bridge the gap between the traditional financial realm and the emerging digital landscape.

Addressing fears and concerns requires knowledge, and education emerged as a powerful tool in empowering individuals and communities. The chapters delving into the importance of education in combatting FUD, fostering regulatory clarity, and promoting financial inclusion emphasized that informed participants are the vanguard of a resilient and thriving fintech ecosystem.

Recognizing the interconnected nature of the cryptoverse, collaboration and government initiatives were highlighted as pivotal. Whether it is collaboration with traditional finance, incentivizing innovation, or fostering a supportive ecosystem through incubators and collaboration hubs, these chapters showcased that a collective effort is essential for the widespread acceptance and integration of fintech solutions.

As we conclude Part IX, we cast our gaze forward with optimism. The challenges outlined in these chapters are not insurmountable obstacles but rather milestones in the evolutionary journey of fintech. Each fear identified is an opportunity for understanding, and each need uncovered is a call for innovation.

Part IX lays the groundwork for what lies ahead — a future where fintech not only addresses fears but becomes an integral part of the global financial landscape. Trust and transparency, cultivated through education, collaboration, and a commitment to user-centric solutions, will be the guiding principles shaping the future of fintech on a global scale.

Part X: Advanced Cryptocurrency- The Summit of Understanding

Introduction to Part X

Welcome to Part X of our exploration into the realms of cryptocurrency evolution, a segment that encapsulates Chapters 90 to 95. As we embark on this journey, we find ourselves navigating the quantum dimensions of a financial landscape continually shaped by innovation, interconnected systems, and the transformative power of blockchain technology.

In this segment, we delve into the intricacies of the cryptoverse, unraveling the complex ecosystem that binds various cryptocurrencies, decentralized exchanges (DEXs), centralized exchanges (CEXs), liquidity pools, and the interconnected nature of digital assets. These chapters serve as a guide through the evolving quantum realms of decentralized finance, where each revelation adds a layer to the understanding of this ecosystem. Chapter 90 lays the foundation by exploring the pivotal role of trust in the cryptoverse. Examining how the cryptoverse needs to build trust through transparency, it establishes the importance of open communication, reliability, and ethical practices in fostering a trustworthy environment.

Chapter 91 delves into the world of markets, scrutinizing how market forces influence the price and demand for cryptocurrencies. Unraveling the complex dance of supply and demand, this chapter sheds light on the quantum interactions that determine the value and adoption of digital assets. Chapter 92 embarks on a comparative exploration of decentralized exchanges (DEXs) and centralized exchanges (CEXs). By dissecting their structures, functionalities, and implications for business adoption in the web3 era, this chapter navigates the contrasting yet interconnected landscapes of these exchange models.

Chapter 94 takes us into the heart of liquidity pools, dissecting the number of tokens available for exchange within organizations. Additionally, it examines asset pools, discussing how blockchain-native coins, fiat currency, or physical/digital assets back cryptocurrencies, and explores the potential for automation through artificial intelligence (AI). In Chapter 95, we broaden our perspective to unveil the interconnectedness of cryptocurrencies as a concept. Analyzing the relationships between diverse digital assets, this chapter paints a holistic picture of the cryptoverse, emphasizing its collective evolution and the interplay of various elements.

As we journey through these chapters, it becomes evident that the cryptoverse is not merely a collection of isolated entities but a ecosystem where quantum entanglements define the course of evolution. The interplay between technology, economics, social s, and market forces paints a picture of a decentralized financial landscape continually in flux.

Join us in this exploration of quantum realms, where each chapter adds a layer of understanding to the ever-evolving narrative of cryptocurrencies. Together, we navigate the complexities, challenges, and opportunities that define the cryptoverse, shaping the future of decentralized finance and digital asset ecosystems.

Chapter 91: Quantum Relationships in Cryptocurrency

Introduction:

In the complex and ever-evolving landscape of cryptocurrency, the entanglement of different digital assets mirrors the quantum entanglement observed in the subatomic realm. Drawing inspiration from Schrödinger's thought experiment and the principles of quantum mechanics, this chapter delves into the entangled nature of cryptocurrencies within the cryptoverse, particularly focusing on the Stablecoin 2.0 model.[466]

The journey begins by exploring the foundational concepts of quantum mechanics, emphasizing the inherent uncertainty and complexity associated with the subatomic world. The analogy between quantum particles and the relationships within the cryptoverse sets the stage for understanding the entanglement of cryptocurrencies.

Stablecoin 1.0, represented by early models like Tether, is examined as the initial attempt to tether digital assets directly to fiat currencies.[467] While providing stability, the limitations and drawbacks of this model become apparent, leading to the emergence of Stablecoin 1.1, as exemplified by XRP. The evolution showcases the cryptoverse's adaptation and pursuit of alternatives beyond traditional fiat entanglements.

---

[466] Smithmyer, Christoher (2023B). "Quantum Entanglement Theory of Crypto Finance." Prerna Foundation at https://medium.com/@prernafoundation/quantum-entanglement-theory-of-crypto-finance-42652d2a73b5. In order to keep this chapter brief, this is a very high level look at the process. If this is an area which interests you I highly recommend you read the source article above.

[467] Smithmyer, Christopher (2023C). The Stablecoin Book. Elite Exclusivity: Detroit.

The exploration extends to Stablecoin 2.0, representing a paradigm shift where entanglements extend beyond mere fiat currencies.[468] The chapter aims to unravel the relationships between different cryptocurrencies within this advanced stablecoin model. Quantum entanglement, as a metaphorical lens, helps elucidate the complexities and interconnectedness inherent in the cryptoverse.

Analytical Approach:

The analytical journey is likened to a concept dropped into a funnel of information, gaining speed as it navigates through various dimensions of quantum mechanics, stablecoin evolution, and the inherent complexities of the cryptoverse.[469] This approach aims to distill a coherent understanding of the entanglement that characterizes the relationships between different cryptocurrencies.

Each time we understand a new link between the cryptoverse's denizens and the physical world, we better understand the markets which affect society. Each time we understand the relationship between different projects within the cryptoverse, we can better understand the movement within the digital realm.[470] Therefore, this area of study is one of the most important areas as the economics of the cryptoverse are being explored.

Fund :

Within the cryptoverse's quantum entanglement, the concept of energy is explored, drawing parallels with quantum reality.[471] Energy, whether internal or externally renewed, is considered vital for the system's functionality. The examination of growth

---

[468] *Id.*

[469] Smithmyer (2023B), *supra*.

[470] *Id.*

[471] *Id.*

without additional energy underscores the inherent s and potential longevity of the cryptoverse. Unto this, the energy entering the cryptoverse is derived from the investments of people living in the physical world.

Alternatively, the movement of the cryptoverse guides the amount of investment it sees by the functionality, market movement, and utility of the systems therein. As such, it becomes a loop in which everything is related, though we do not understand the relationships as of yet.[472]

Conclusion:

The chapter concludes by highlighting the growth potential within the cryptoverse's quantum entanglement. Acknowledging the distortions and complexities introduced at different levels of evolution, the analysis emphasizes the adaptability of the system. The analogy of crypto's entanglement with quantum mechanics serves as a conceptual framework, offering insights into the nuanced relationships and interactions within the digital financial realm.

As the study unfolds, it becomes evident that while the cryptoverse may share metaphorical entanglements with quantum particles, its s are shaped by economic theories, social interactions, and technological advancements. The entangled cryptoverse remains a frontier where understanding and managing the inherent complexities contribute to its growth and resilience.

---

[472] Smithmyer, (2023C), *supra*.

Chapter 92: Market Dynamics in Cryptocurrencies

Introduction:

The world of cryptocurrencies is characterized by its   and volatile nature, where market forces play a pivotal role in shaping the prices and demand for digital assets. Chapter 92 delves into the relationship between market s and the cryptocurrency landscape. Understanding the factors that influence these s is essential for investors, traders, and enthusiasts seeking to navigate the ever-changing crypto markets.

Market Dynamics in Cryptocurrencies:

- **Supply and Demand:** At the core of markets lies the fundamental economic principle of supply and demand. The fixed or capped supply of many cryptocurrencies, such as Bitcoin with its 21 million cap, creates a scenario where scarcity can drive up demand.[473] When demand surpasses supply, prices tend to rise, and conversely, when demand diminishes, prices may fall.[474]
- **Market Sentiment:** Cryptocurrency markets are heavily influenced by sentiment. Positive news, regulatory developments, or technological advancements can create bullish sentiment, prompting increased buying

---

[473] Staff (2023). "Crypto Trading 101: Crypto Trader's Guide to Supply and Demand Trading." Cryptohopper at https://www.cryptohopper.com/blog/crypto-trader-s-guide-to-supply-and-demand-trading-7771.

[474] This is the general understanding of supply and demand in cryptocurrency; however, if there is not an incentive for the currency to go up in value, reduction in supply does not matter if there is not sufficient demand.

activity.[475] On the contrary, negative news, regulatory uncertainties, or security breaches can induce bearish sentiment, leading to sell-offs and price declines.

- **Regulatory Environment:** Regulatory developments and government interventions have a profound impact on cryptocurrency markets. Clarity and favorable regulations can boost investor confidence, leading to increased demand.[476] Conversely, regulatory uncertainties or strict measures can lead to a decline in demand as investors may perceive higher risks.
- **Technological Advancements:** Innovations and technological upgrades within the crypto space can influence markets.[477] Upgrades to blockchain networks, the introduction of new consensus mechanisms, or the development of interoperability solutions can enhance the utility and adoption of specific cryptocurrencies, thereby impacting their prices.[478]
- **Market Liquidity:** Liquidity, or the ease with which an asset can be bought or sold without affecting its price, plays a crucial role.[479] Cryptocurrencies with higher liquidity are often more attractive to traders and investors. Illiquid markets may experience more

---

[475] Staff (2023). "Crypto Trading 101: What is Market Sentiment." Cryptohopper at https://www.cryptohopper.com/blog/what-is-crypto-market-sentiment-and-why-does-it-matter-4198.

[476] Dewey, Josias & Patel, Samir (2024). "Blockchain and Cryptocurrency Laws and Regulations 2024 USA." Holland and Knight at https://www.globallegalinsights.com/practice-areas/blockchain-laws-and-regulations/usa.

[477] Staff (2023). "The impact of Cryptocurrency on the Tech Industry." Solu Labs at https://www.solulab.com/cryptocurrency-tech-industry-impact/.

[478] Staff (2023). "What is Crypto Liquidity and How does it impact the market?" B2Broker at https://b2broker.com/news/what-is-crypto-liquidity-and-how-does-it-impact-market-stability/.

[479] *Id.*

significant price fluctuations due to larger spreads between buy and sell orders.

- **Market Manipulation:** Cryptocurrency markets, particularly those with lower market capitalization, are susceptible to manipulation.[480] Whales, or large holders of cryptocurrencies, can influence prices by executing substantial trades. Pump-and-dump schemes, where prices are artificially inflated and then deliberately crashed, are examples of manipulation that impact market dynamics.[481]
- **Integration with Traditional Finance:** As cryptocurrencies become more integrated into traditional financial systems, their s are increasingly influenced by traditional market factors. Economic indicators, interest rates, and geopolitical events in the global financial landscape can spill over into the cryptocurrency markets.[482]
- **Technological and Security Concerns:** The security of blockchain networks and the technology underpinning cryptocurrencies is paramount.[483] Incidents of hacking, vulnerabilities, or flaws in the underlying technology can erode investor confidence, leading to sell-offs.

---

[480] Sing, Onkar (2023). "What is market manipulation in cryptocurrency?" Coin Telegraph at https://cointelegraph.com/explained/what-is-market-manipulation-in-cryptocurrency.

[481] Nelson, Danny (2024). "Crypto Boosters Attack SEC for 'Manipulating' BTC Market after ETF Tweet." Coin Desk at https://www.coindesk.com/policy/2024/01/10/crypto-boosters-attack-sec-for-manipulating-btc-market-after-etf-tweet/.

[482] Arasaratnam, Mark & Solot, Ilan (2023). "The handover begins: TradFi takes center stage in Crypto's Next Phase." Coin Desk at https://www.coindesk.com/consensus-magazine/2023/11/07/the-handover-begins-tradfi-takes-center-stage-in-cryptos-next-phase/.

[483] Staff (2022). "Is Cryptocurrency Safe? Things to Be Aware of." N26 at https://n26.com/en-at/blog/is-cryptocurrency-safe.

Conversely, advancements in security measures can enhance trust and bolster demand.

Conclusion:

The cryptocurrency market is an ecosystem where various factors converge to shape prices and demand. Understanding these markets is crucial for participants looking to navigate the complexities of the crypto landscape. From the basic principles of supply and demand to the nuanced impact of regulatory decisions and technological advancements, each element plays a role in influencing market behavior.

As the crypto space continues to evolve, market participants must remain vigilant, staying informed about the latest developments and adapting strategies to navigate the ever-changing landscape. The interconnectedness of global financial systems, technological innovations, and regulatory environments underscores the need for a holistic approach when analyzing markets.

In summation, while cryptocurrencies offer unique opportunities for growth and innovation, they also present challenges and risks inherent in a nascent and market. This chapter provides a comprehensive exploration of these s, shedding light on the multifaceted factors that contribute to the ebb and flow of prices and demand within the cryptocurrency space.

Chapter 93: Decentralized Exchanges (DEXs) vs. Centralized Exchanges (CEXs)

Introduction:

CEX- Centralized Exchange
DEX- Decentralized Exchange

The advent of blockchain technology and cryptocurrencies has led to the emergence of decentralized finance (DeFi) platforms, with Decentralized Exchanges (DEXs) and Centralized Exchanges (CEXs) being two primary models for trading digital assets. In the context of business adoption of Web3 and cryptocurrency, the choice between DEXs and CEXs plays a pivotal role in determining the efficiency, security, and regulatory compliance of trading operations. This chapter explores and contrasts DEXs and CEXs, shedding light on their distinct features, advantages, challenges, and implications for businesses entering the Web3 and cryptocurrency space.

Decentralized Exchanges (DEXs):

Definition and Functionality:

- DEXs operate on decentralized networks and smart contracts, enabling users to trade directly with one another without the need for intermediaries.[484]
- Smart contracts facilitate the execution of trades by automating order matching, ensuring transparency and immutability of transactions.

Trustless Transactions:

---

[484] Staff (2023). "Centralized Exchanges (CEXs) and Decentralized Exchanges (DEXs) Which is the best crypto exchange?" Centric at https://medium.com/@centicio/centralized-exchange-cex-vs-decentralized-exchange-dex-which-is-the-best-crypto-exchange-148f48ea51c1.

- DEXs are designed to be trustless, meaning users retain control of their private keys and funds throughout the trading process.
- This eliminates the need for users to deposit funds into a centralized entity, reducing counterparty risk.[485]

Censorship Resistance:

- DEXs often leverage blockchain technology, providing censorship resistance and ensuring that users can access and trade assets without restrictions.[486]

Asset Variety:

- DEXs typically support a broader range of assets, including those issued on various blockchain networks, promoting interoperability within the decentralized ecosystem.[487]

Global Accessibility:

- Users from anywhere in the world can access DEXs, fostering inclusivity and democratizing access to financial services.

Liquidity Pools:

- Liquidity on DEXs is often provided by users who contribute to liquidity pools, earning fees in return. This decentralized liquidity model contrasts with CEXs where the exchange manages liquidity.[488]

---

[485] Smithmyer (2023) *supra*.
[486] *Id*.
[487] Centric *supra*.
[488] *Id*.

Centralized Exchanges (CEXs):

Centralized Control:

- CEXs are operated and managed by centralized entities that act as intermediaries facilitating trades on behalf of users.
- Users must deposit funds into exchange-controlled wallets, relinquishing a degree of control over their assets.[489]

Order Matching and Liquidity:

- CEXs manage order matching centrally, providing high-speed and efficient trading services.
- Liquidity is often managed by the exchange itself, ensuring a seamless trading experience for users.

Regulatory Compliance:

- CEXs are subject to regulatory frameworks requiring compliance with Know Your Customer (KYC) and Anti-Money Laundering (AML) regulations.
- This regulatory adherence can instill confidence in users and attract institutional participants seeking a compliant trading environment.[490]

User Experience:

- CEXs typically offer a user-friendly interface and a familiar trading experience, making them accessible to users with varying levels of technical expertise.

Security Concerns:

---

[489] Smithmyer (2023), *supra*.
[490] *Id.*

- While CEXs implement security measures, they are susceptible to centralized points of failure. Successful cyberattacks on CEXs have resulted in significant losses.[491]

Comparison:

Side by side, CEXs and DEXs offer different advantages and drawbacks to members of the digital community.

- Security:
    a. DEXs, by design, offer enhanced security through trustless, blockchain-based transactions. Users maintain control of their private keys, reducing the risk of hacks or mismanagement.
    b. CEXs, despite implementing security protocols, are susceptible to targeted attacks due to their centralized nature. The custody of user funds by the exchange introduces a potential single point of failure.
- User Privacy:
    a. DEXs prioritize user privacy by allowing users to trade without the need for KYC procedures. This attracts users to value anonymity in their financial transactions.
    b. CEXs, due to regulatory compliance requirements, often mandate KYC processes, compromising user privacy for the sake of legal adherence.
- Speed and Efficiency:
    a. CEXs typically offer faster transaction speeds and order matching capabilities since these

---

[491] Centric, *supra*.

processes are centralized and managed by the exchange.

b. DEXs may face challenges related to latency and speed due to the decentralized nature of blockchain networks and reliance on smart contracts.

- Regulatory Compliance:
  a. CEXs adhere to regulatory standards, making them suitable for institutional investors and businesses that require compliance with legal frameworks.
  b. DEXs, operating in a more decentralized manner, may face regulatory uncertainties. However, efforts are being made to integrate compliance features without compromising decentralization.
- Ownership and Control:
  a. DEXs emphasize user ownership and control over assets. Users retain control of their private keys, reducing reliance on third-party entities.
  b. CEXs require users to deposit funds into exchange-controlled wallets, leading to a degree of reliance on the centralized entity. Users trust the exchange to safeguard their assets.

Conclusion:

In the area Web3 and cryptocurrency adoption, the choice between DEXs and CEXs depends on various factors, including user preferences, security considerations, and regulatory requirements. DEXs offer a trustless and more decentralized system while CEXs are more controlled and have a higher volume. Both have a strong place in the cryptoverse to secure a future where individuals will be able to access their wealth on their terms ensuring a more inclusive cryptoverse.

Chapter 94: Liquidity Pools, Asset Pools, and Automated Pools

Introduction:

In the world of cryptocurrency and decentralized finance (DeFi), liquidity pools and asset pools play pivotal roles in facilitating efficient trading and ensuring the stability of blockchain-based assets. Chapter 94 delves into the mechanisms of liquidity pools, examining their significance in both decentralized and centralized exchanges. Furthermore, it explores the concept of asset pools, elucidating how they serve as the backbone for various digital currencies and how advancements in artificial intelligence (AI) have introduced automation to enhance the management of these pools. In interest of space, this section is broken down into bullet lists.

Liquidity Pools in Depth:

- Definition and Function:
    - *Liquidity pools* refer to reserves of tokens locked in smart contracts, enabling users to trade on decentralized exchanges (DEXs) without relying on traditional order book models. Users contribute their tokens to these pools, and, in return, they receive liquidity provider (LP) tokens representing their stake.[492] It also functions to mean the reserves which allow for sales of tokens.
- Automated Market Makers (AMMs):
    - Liquidity pools operate on the principle of AMMs, a decentralized pricing mechanism. Popularized by platforms like Uniswap, AMMs

---

[492] Rouse, Margaret (2023). "Liquidity Pool." Techopedia at https://www.techopedia.com/definition/liquidity-pool-defi-liquidity-pool.

use mathematical formulas to determine token prices based on the ratio of assets in the pool.[493] This algorithmic approach eliminates the need for a traditional order book.

- Balancing Supply and Demand:
    - The liquidity in a pool ensures that there are enough tokens available for exchange, mitigating slippage and providing users with competitive prices. As demand for a particular token increases, arbitrageurs can capitalize on price discrepancies by trading against the pool and, in doing so, help restore equilibrium.[494]
- Impermanent Loss:
    - Liquidity providers expose themselves to impermanent loss, a phenomenon where the value of their assets diverges from holding them in a wallet.[495] This risk arises due to price fluctuations and is a consideration for those participating in liquidity provision.
- DEXs and Centralized Exchanges (CEXs):
    - In the context of centralized exchanges, ensuring there is enough liquidity for trading pairs is crucial. Centralized exchanges maintain liquidity by having reserves of tokens available for immediate execution of trades.[496] The comparison between liquidity pools on DEXs

---

[493] Staff (2023). "What are Automated Market Makers AMMs?" Chainlink at https://chain.link/education-hub/what-is-an-automated-market-maker-amm.
[494] Cryptohopper, *supra.*
[495] Staff (2023). "What is impermanent loss? How to manage it in DeFi liquidity pools." Crypto.com at https://crypto.com/university/what-is-impermanent-loss.
[496] Centric, *supra.*

and reserves on CEXs highlights the differences in operational models.

Asset Pools:

- Defining Asset Pools:
  - *Asset pools* refer to the collection of assets, which can include blockchain-native coins, fiat currency, or physical/digital assets, backing a specific cryptocurrency.[497] These pools provide a foundation for the value and stability of the associated digital currency.
- Blockchain-Native Coins:
  - Many cryptocurrencies have their own blockchain-native coins, and the value of these coins is often supported by the community, consensus mechanisms, and the underlying technology. These assets contribute to the overall value of the asset pool.
- Fiat Currency Backing:
  - Some stablecoins are backed by fiat currencies, such as the US Dollar or Euro. Asset pools for these stablecoins hold reserves in the corresponding fiat currency, ensuring that the digital currency maintains a stable and pegged value.[498]
- Physical/Digital Asset Backing:
  - Asset pools can also include physical or digital assets that serve as collateral. In decentralized finance, protocols often use collateralized assets to issue synthetic versions of traditional assets, expanding the range of assets in the pool.[499]

---

[497] Smithmyer (2023A), *infra*.
[498] Smithmyer (2023), *supra*.
[499] Smithmyer (2023A), *infra*.

- Stability and Value Proposition:
    - The stability and value of a digital currency are directly linked to the assets held in its pool. Transparent management and adequate reserves contribute to the credibility and trustworthiness of the currency, attracting users and investors.[500]

Automated Market Maker: A pool where users can deposit currency to create liquidity.

Automating Pools with AI:

- Integration of Artificial Intelligence:
    - The integration of AI in managing liquidity and asset pools introduces efficiency and responsiveness. AI algorithms can analyze market conditions, predict trends, and automate strategies to optimize the composition of these pools.[501]
- Automated Rebalancing:
    - AI algorithms can continuously monitor and analyze the composition of liquidity pools, automatically rebalancing assets to adapt to changing market conditions.[502] This ensures that

---

[500] *Id.*

[501] Acceswire (2021). "AI-based trading pool KB going live, Assures minimum 5-15% ROI every month and automated payments." Yahoo Finance at https://finance.yahoo.com/news/ai-based-trading-pool-kb-204500469.html.

[502] Robin, Konstantin (2023). "AI-Powered Portfolio Management in Crypto-Optimized Diversification for Long-term success." The Daily Hodl at https://dailyhodl.com/2023/11/09/ai-powered-portfolio-management-in-crypto-optimizing-diversification-for-long-term-success/.

the pool remains optimized for liquidity provision.

- Predictive Analysis for Liquidity:
    - AI tools can utilize predictive analytics to anticipate changes in liquidity demand. By analyzing historical data and market trends, these tools can optimize liquidity provision in advance, reducing the impact of sudden fluctuations.[503]
- Risk Management:
    - AI-driven risk management tools can assess potential impermanent loss and ally adjust the allocation of assets in liquidity pools.[504] This proactive risk mitigation strategy enhances the overall resilience of liquidity provision.
- Enhanced Decision-Making:
    - AI provides decision-making support by processing vast amounts of data in real-time. This includes factors such as market sentiment, trading volumes, and external events, enabling more informed and data-driven decisions regarding liquidity and asset pool management.[505]

---

[503] Staff (2023). "Machine Learning and Predictive Analytics in crypto liquidity aggregation: A game changer for traders." Fluid AI at https://fluidai.com/blog/machine-learning-and-predictive-analytics-in-crypto-liquidity-aggregation-a-game-changer-for-traders.

[504] Staff (2023). "Crypto Trading Risk Management: Navigating the Volatile Crypto Markets." Coin DCX at https://coindcx.com/blog/cryptocurrency/crypto-trading-risk-management-for-beginners/.

[505] Nemanis, Mantas (2023). "The role of artificial intelligence in crypto trading: Enhanced Decision Making." Linked in at https://www.linkedin.com/pulse/role-artificial-intelligence-crypto-trading-enhanced-mantas-nemanis-haq7e/.

Conclusion:

In conclusion, liquidity pools and asset pools form the backbone of decentralized and centralized exchanges, providing the liquidity necessary for the seamless operation of cryptocurrency markets. The innovative concepts of liquidity pools, especially in the context of decentralized exchanges, have reshaped how users engage in trading activities. Asset pools, encompassing a diverse range of assets, are instrumental in establishing the stability and value proposition of digital currencies.

As we look to the future, the role of artificial intelligence in automating the management of these pools is becoming increasingly significant. AI brings precision, predictive capabilities, and efficiency to the landscape of decentralized finance, ensuring that liquidity and asset pools can adapt to the ever-changing conditions of the cryptocurrency market. By combining the principles of liquidity provision, asset backing, and AI-driven automation, the cryptocurrency ecosystem continues to evolve, offering enhanced opportunities for users, investors, and the broader financial landscape.

Chapter 95: Interconnectedness of Cryptocurrencies (Non-Quantum)

Introduction:

Chapter 95 explores the web of interconnectedness that defines the world of cryptocurrencies. In this and evolving landscape, individual digital currencies are not isolated entities but rather nodes in a complex network. The interconnectedness of cryptocurrencies goes beyond mere transactions; it encompasses technological, economic, and social dimensions. This chapter delves into the various facets of this interconnected ecosystem, shedding light on how different cryptocurrencies influence and rely on each other, ultimately shaping the collective future of decentralized finance.

Interconnected Technological Foundations:

- Blockchain as the Common Thread:
    - At the heart of the interconnectedness of cryptocurrencies lies blockchain technology. Most cryptocurrencies leverage some form of blockchain, serving as a shared foundation for transactions, security, and decentralization.[506] Bitcoin, as the pioneering cryptocurrency, laid the groundwork for subsequent projects to build upon.
- Forking and Codebase Relationships:
    - The open-source nature of many blockchain projects has led to the forking of codebases. Forks, whether soft or hard, create new cryptocurrencies with shared ancestry.

---

[506] Sharma (2023A). "Is there a cryptocurrency price correlation to the stock market." Investopedia at https://www.investopedia.com/news/are-bitcoin-price-and-equity-markets-returns-correlated/.

Ethereum, for instance, has spawned various tokens and projects through its smart contract capabilities, contributing to the interconnected Ethereum ecosystem.

- Interoperability Protocols:
    - Projects aiming to enhance interoperability between different blockchains have emerged. Platforms like Polkadot and Cosmos facilitate communication and asset transfer between disparate blockchains.[507] This interoperability fosters a collaborative environment, allowing different cryptocurrencies to complement each other's strengths.

Economic Interdependencies:

- Bitcoin as Digital Gold:
    - Bitcoin, often referred to as digital gold, establishes a foundational economic interdependence within the cryptocurrency space. Its role as a store of value influences the market sentiment and investment strategies of other cryptocurrencies. Movements in the price and adoption of Bitcoin have a ripple effect across the entire market.
- Altcoins and Tokenomics:
    - The proliferation of alternative coins (altcoins) and tokens introduces diverse economic models. These digital assets often serve specific functions within decentralized applications or ecosystems. The success or failure of these projects can influence the broader cryptocurrency market, creating economic interdependencies.

---

[507] Cryptoslate at https://cryptoslate.com/cryptos/interoperability/.

- Decentralized Finance (DeFi) Ecosystem:
    - The rise of decentralized finance further accentuates the interconnectedness of cryptocurrencies. DeFi projects often collaborate, share liquidity, and utilize each other's services. A change or innovation in one DeFi protocol can have cascading effects, influencing the strategies and development approaches of others.

Social s and Community Influence:

- Cross-Pollination of Communities:
    - Cryptocurrency communities play a vital role in shaping the narrative and direction of projects. The interconnectedness of these communities fosters cross-pollination of ideas. Members of one cryptocurrency community may actively participate or invest in other projects, creating a shared sense of innovation and progress.
- Social Media and Information Flow:
    - Social media platforms serve as hubs for discussions, news dissemination, and community engagement. Information, whether positive or negative, quickly spreads across different cryptocurrency communities, impacting market sentiments and influencing investor decisions. This interconnected information flow contributes to the correlated nature of the market.
- Collective Response to Challenges:
    - When faced with challenges or external threats, the cryptocurrency community often rallies collectively. Market-wide responses to regulatory developments, security incidents, or

technological advancements highlight the shared destiny of cryptocurrencies. The interconnectedness of these responses reflects the unified stance of the broader ecosystem.

Challenges and Opportunities:

- Market Volatility and Correlation:
    - The interconnected nature of cryptocurrencies contributes to market volatility and correlation. Events affecting one major cryptocurrency can trigger cascading effects, leading to market-wide movements. While this interconnectedness poses challenges, it also presents opportunities for diversified portfolios and risk management strategies.
- Innovation and Synergy:
    - Interconnectedness fosters innovation through collaborative efforts. Cryptocurrency projects often learn from each other's successes and failures, driving continuous improvement. Synergies emerge as projects integrate functionalities from different blockchains, creating a richer and more interconnected technological landscape.

Conclusion:

In conclusion, the interconnectedness of cryptocurrencies transcends mere financial transactions; it encompasses shared technological foundations, economic interdependencies, and vibrant social s. The evolution of blockchain technology and the proliferation of diverse digital assets contribute to a complex yet synergistic ecosystem. The economic success or challenges faced by individual cryptocurrencies reverberate across the entire landscape, shaping the collective narrative of decentralized finance.

As the cryptocurrency space continues to mature, recognizing and understanding this interconnectedness becomes paramount. Market participants, developers, and enthusiasts alike navigate a landscape where the fate of one cryptocurrency is tied to the fortunes of others. Embracing this interconnected reality opens avenues for collaborative innovation, risk mitigation, and the collective pursuit of a decentralized future that transcends individual projects and resonates across the broader cryptocurrency ecosystem.

Conclusion Part X

As we conclude Part X, which delved into the multifaceted dimensions of cryptocurrency evolution, it is evident that we are navigating the quantum realms of a transformative financial landscape. The chapters explored themes ranging from the relationships between cryptocurrencies to the technological foundations that bind them, delving into the markets, the dichotomy of decentralized and centralized exchanges, and the liquidity pools that underpin the entire ecosystem. Together, these chapters form a mosaic, illustrating the interconnected and ever-evolving nature of the cryptoverse.

The chapters underscored the interconnectedness of cryptocurrencies as the quantum glue that binds this decentralized realm. From the shared technological foundations of blockchain to economic interdependencies and social s, the cryptoverse operates as a and entangled network. This interconnectedness is not merely a matter of transactions but extends to the shared ethos, technology, and economic underpinnings that define the collective evolution of cryptocurrencies.

At the core of this interconnected realm lies the revolutionary technology of blockchain. The chapters explored how blockchain serves as a common thread, enabling the creation of diverse cryptocurrencies and projects. Forking, interoperability protocols, and the evolution of codebases contribute to a shared quantum space where innovation in one corner of the cryptoverse reverberates across the entire ecosystem. Economically, the quantum relationship between cryptocurrencies was explored through the lens of market s and the role of major players like Bitcoin. The rise of alternative coins and decentralized finance (DeFi) further demonstrated the complex economic interdependencies at play. The chapters emphasized how developments in one corner of the cryptoverse have cascading effects, influencing market sentiment and strategies across the spectrum.

Beyond technology and economics, the social s within the cryptoverse were unraveled. The interconnectedness of communities, cross-pollination of ideas, and the collective response to challenges highlighted the shared destiny and collaborative spirit that define the quantum social fabric of cryptocurrencies. Social media platforms emerged as conduits for information flow, impacting market sentiments and shaping the narrative of the cryptoverse. Acknowledging the challenges posed by market volatility and correlation, the chapters also illuminated the opportunities embedded in this interconnected landscape. The ability to learn from each other's successes and failures, coupled with the synergies arising from collaborative innovation, positions the cryptoverse at the forefront of a quantum financial revolution.

As we navigate the quantum realms of cryptocurrency evolution, it becomes evident that the journey is ongoing. The cryptoverse is not a static entity but a , ever-shifting landscape where quantum entanglements define the trajectory. The challenges posed by market volatility, regulatory uncertainties, and technological innovations create a vibrant canvas for exploration and adaptation.

In summary, Part X invites participants, observers, and enthusiasts to embrace the quantum future of the cryptoverse. The interconnectedness explored in these chapters is not a hindrance but a catalyst for collective growth. As we venture further into the uncharted territories of decentralized finance, quantum economics, and technological innovation, it is the interconnected nature of cryptocurrencies that propels us forward, ensuring that the quantum realms we navigate are not just complex but infinitely rich with potential.

Part XI: NFTs- The Frontiers of the Digital Realm

Introduction to Part XI:

In the ever-evolving landscape of digital innovation, Part XI of our exploration embarks on a journey into the forefront of blockchain technology, shining a spotlight on Non-Fungible Tokens (NFTs) and the realm of smart contracts. As we step into this segment, we find ourselves at the intersection of creativity and programmability, where decentralized technologies are reshaping the way we perceive and interact with digital assets.

Chapters 96 to 100 serve as a compass, guiding us through the multifaceted landscape of NFTs—unique cryptographic tokens that have transcended the boundaries of art and evolved into powerful tools for representation, memorialization, and contractual agreements. Each chapter unfolds a layer of the NFT narrative, revealing its potential to redefine ownership, authenticity, and the very nature of digital value.

At the core of this exploration lies the symbiotic relationship between NFTs and smart contracts. Smart contracts, the programmable self-executing agreements on the blockchain, emerge as the architectural foundation upon which the uniqueness and functionality of NFTs are built. This integration not only enhances the capabilities of digital assets but also introduces a paradigm shift in how we conceptualize and execute agreements in the digital realm.

As we venture into the chapters ahead, themes of decentralization, trustless execution, and the tokenization of assets will be recurrent motifs. The exploration of NFTs as digital receipts, representations, memorials, and smart contracts will unveil the diverse trajectories these tokens can take, providing a glimpse into the expansive future of blockchain technology.

While the potential of NFTs and smart contracts is vast, it is essential to navigate the challenges and considerations that accompany these innovations. From environmental concerns to inclusivity in adoption, each chapter delves into the complexities of this transformative landscape, urging us to tread carefully as we navigate the unknown.

Part XI is an invitation to envision a future where digital assets are not only programmable but also uniquely representative and seamlessly integrated into the fabric of our digital lives. It encourages us to explore the intersections of technology and creativity, and to actively contribute to the narrative of tomorrow.

As we embark on this exploration into the frontiers of digital transformation, let us embrace the unknown with curiosity, navigate the complexities with diligence, and craft a narrative where innovation converges with imagination to shape a decentralized future. Exploration is the heart of business; while wages are forged in the fires of the known world, fortunes are built in the exploration of new frontiers.

Chapter 96: Introduction to NFTs

Introduction

As we explore digital assets in the cryptoverse, a revolutionary concept has emerged, challenging traditional notions of ownership, provenance, and value in the digital realm. Welcome to the world of Non-Fungible Tokens (NFTs), a paradigm-shifting phenomenon that has captured the imagination of artists, creators, collectors, and investors alike.

NFTs represent a groundbreaking fusion of blockchain technology, digital art, and tokenization, offering a new dimension to ownership and authenticity in the digital space. In this chapter, we embark on a journey into the intricacies of NFTs, exploring their origins, underlying technology, diverse applications, and the profound impact they've had on the art, entertainment, and gaming industries.

Understanding NFTs: Unveiling the Digital World

At its core, an NFT is a unique digital asset that is indivisible, verifiable, and non-interchangeable, setting it apart from fungible cryptocurrencies like Bitcoin or Ethereum. The uniqueness of NFTs is facilitated by blockchain technology, which provides a transparent and immutable ledger, ensuring the provenance and scarcity of each token.[508]

The concept of NFTs gained prominence with the Ethereum blockchain, leveraging its smart contract capabilities to enable the creation and management of these distinctive digital assets. Ethereum's ERC-721 standard, specifically designed for non-fungible tokens, laid the groundwork for the explosion of creativity and innovation in the NFT space.[509]

---

[508] Sharma (2023), *supra*.
[509] Id.

Origins and Evolution: From CryptoKitties to Digital Masterpieces

To truly appreciate the impact of NFTs, we must delve into their origins. The NFT craze first captured public attention with CryptoKitties, a blockchain-based game where users could buy, sell, and breed unique virtual cats.[510] This experiment showcased the potential for scarce digital assets and paved the way for more sophisticated applications.

The real breakthrough, however, came when artists and creators recognized the transformative power of NFTs in the art world. Digital artists could now tokenize their work, providing a secure and transparent way to establish ownership and provenance. This paradigm shift challenged traditional art markets, offering a direct relationship between creators and consumers while ensuring artists received fair compensation for their digital creations.

Applications Beyond Art: NFTs in Gaming, Music, and Real Estate

As NFTs gained momentum, their applications extended far beyond the realm of digital art. The gaming industry embraced NFTs, introducing unique in-game assets, characters, and items that players could truly own and trade. This not only added a new layer of authenticity to virtual worlds but also created economic ecosystems within games.[511]

In the music industry, NFTs opened avenues for artists to tokenize their music, concert tickets, and even royalties. Fans could now own a piece of their favorite artist's work, fostering a deeper connection between creators and their audience. Additionally, NFTs offered solutions to longstanding challenges

---

[510] CryptoKitties at https://www.cryptokitties.co/.
[511] Smithmyer et at (2022) *supra*.

in royalty distribution, ensuring artists received their fair share directly and transparently.

The concept of tokenization also found its way into the real estate sector, where NFTs represented ownership or fractional ownership of physical properties. This innovative approach promised to streamline property transactions, enhance liquidity, and democratize access to real estate investments.

While the rise of NFTs brought about unprecedented opportunities, it also sparked debates and criticisms.[512] Environmental concerns related to the energy consumption of blockchain networks hosting NFTs, issues of copyright infringement, and market speculation leading to inflated prices are among the challenges that have emerged.

Critics argue that the NFT space is experiencing a speculative bubble, drawing parallels to historical market bubbles. Additionally, concerns about the environmental impact of energy-intensive proof-of-work blockchains have prompted discussions about transitioning to more sustainable alternatives.

Conclusion

In conclusion, Non-Fungible Tokens represent a pivotal moment in the ongoing digital renaissance. As we navigate the landscape of NFTs, we witness the convergence of technology, art, and economics, redefining how we perceive and interact with digital assets.

The transformative power of NFTs extends beyond a mere market trend; it signifies a shift in paradigms, challenging traditional notions of ownership, authenticity, and creativity. From digital art and gaming to music and real estate, NFTs have

---

[512] Staff (2023). A Beginners guide on the legal risks and issues around NFTs." Coin Telegraph at https://cointelegraph.com/learn/a-beginners-guide-on-the-legal-risks-and-issues-around-nfts.

woven themselves into the fabric of various industries, leaving an indelible mark on the digital landscape.

Join us in this exploration of NFTs, where we unravel the threads of innovation, creativity, and technological advancement that compose the ecosystem of this digital revolution. As we delve deeper into the applications, challenges, and implications of NFTs, we gain insights into a future where the boundaries between the physical and digital worlds continue to blur, and ownership takes on a new and exciting form.

Chapter 97: NFTs as Digital Receipts

Introduction

In the landscape of online transactions, a revolutionary concept has emerged - Non-Fungible Tokens (NFTs) serving as digital receipts. Traditionally, receipts were mundane records of financial transactions, often printed on paper or sent through emails. However, the integration of NFTs has introduced a paradigm shift, transforming these transaction records into unique, verifiable, and valuable digital assets.

Non-Fungible Token- unique tokens on a blockchain which are not replicated, each is an individual and can be identified.

This chapter delves into the innovative use of NFTs as digital receipts, exploring the implications for businesses, consumers, and the broader landscape of online commerce.

The Evolution of Digital Receipts: From Mundane to Unique

Digital receipts have long been a staple in e-commerce, providing consumers with records of their transactions. However, these receipts were often generic, lacking the uniqueness and authenticity associated with physical receipts.[513] With the introduction of NFTs, digital receipts are undergoing a metamorphosis, becoming personalized, secure, and tradable assets.

---

513 Cramer, Peter & O'Rourke Brendan (2023). "As NEFTs blur the line between 'reciept and 'product', trademark owners fight over the new virtual markets." JDsupra at https://www.jdsupra.com/legalnews/as-nfts-blur-the-line-between-receipt-7593211/.

NFTs as digital receipts leverage blockchain technology to imbue each receipt with a distinct cryptographic signature. This not only ensures the authenticity and integrity of the transaction record but also introduces an element of ownership and rarity. Each digital receipt, represented as an NFT, becomes a unique asset that consumers can truly own, trade, or showcase as proof of their digital interactions.[514]

Benefits for Businesses: Enhancing Transparency and Customer Engagement

For businesses, the adoption of NFTs as digital receipts brings a myriad of advantages. Firstly, the transparency embedded in blockchain technology ensures an unalterable record of transactions. This transparency not only mitigates the risk of fraudulent activities but also builds trust between businesses and their customers.[515]

Moreover, the unique and tradable nature of NFT receipts opens up avenues for enhanced customer engagement. Businesses can incentivize customer loyalty by offering exclusive NFTs tied to their purchases. These digital assets, whether they carry virtual collectibles, discounts, or access to premium content, create an interactive and rewarding experience for consumers.

Empowering Consumers: Ownership and Verification

From a consumer standpoint, the adoption of NFTs as digital receipts marks a departure from the ephemeral nature of conventional transaction records. Owning an NFT receipt means possessing a digital asset with intrinsic value. This ownership element contributes to a sense of exclusivity and permanence,

---

[514] *Id.*

[515] Bussler, Frederik (2022). "Why NFTs are the Future of Invoicing." Hackernoon at https://hackernoon.com/why-nfts-are-the-future-of-invoicing.

as opposed to conventional receipts that can easily be deleted or lost.[516]

NFT receipts also offer a novel way for consumers to verify their transaction history. By having a unique cryptographic token associated with each purchase, consumers can independently verify the authenticity of their receipts, reducing reliance on centralized systems. This empowerment aligns with the broader ethos of decentralization and user autonomy that blockchain technology champions.

Challenges and Considerations: Scalability and Environmental Impact

While the concept of NFTs as digital receipts brings forth numerous benefits, it is not without challenges. One notable concern is scalability, especially as the popularity of NFTs continues to rise. Blockchain networks, particularly those utilizing proof-of-work consensus mechanisms, face challenges in handling the growing demand for transaction processing. This raises questions about the scalability and efficiency of NFTs as a solution for widespread use in digital receipts.

Environmental considerations related to the energy consumption of blockchain networks also emerge as a critical consideration. As the debate surrounding the environmental impact of blockchain intensifies, finding sustainable solutions becomes imperative to ensure the long-term viability of NFTs.

Conclusion:

The integration of NFTs as digital receipts represents a transformative leap in the realm of online transactions. Beyond serving as mere records, digital receipts, as NFTs, become unique, tradable assets that enhance transparency, empower consumers, and foster engagement.

---

[516] Cramer & O'Rourke, *supra*.

As businesses explore the potential of this innovative approach, addressing scalability challenges and embracing sustainable blockchain solutions will be crucial. NFTs as digital receipts herald a new era where the mundane becomes extraordinary, and each transaction leaves a lasting digital imprint, reinforcing the profound impact of blockchain technology on the way we conduct and perceive online commerce.

Chapter 98: NFTs as Representations of Items (Titles)

Introduction:

In the expansive realm of digital ownership and representation, Non-Fungible Tokens (NFTs) have emerged as the driving force behind a transformative renaissance. Beyond their role as digital assets, NFTs are now reshaping the concept of titles and how items are represented in both virtual and tangible spaces. This chapter delves into the revolutionary impact of NFTs on titles and item representation, exploring the profound implications for industries ranging from gaming to real estate.

Unveiling the Power of NFTs in Titles: From Digital Collectibles to Real-world Assets

Traditionally, titles and ownership documentation have been associated with physical certificates or digital records stored in centralized databases. The advent of NFTs introduces a paradigm shift by providing a decentralized and verifiable means of establishing ownership.[517] NFTs serve as digital titles, uniquely representing ownership rights and capturing the essence of tangible or digital items.

In the digital realm, this is prominently observed in the world of gaming. Gamers can now truly own in-game assets through NFTs, obtaining a digital title that is securely stored on the blockchain. These digital titles extend beyond the virtual space, impacting how real-world assets are represented and transferred.

---

[517] Bernstein, Corinne (2023). "Top 7 NFT Use Cases forBusiness." Tech Target at https://www.techtarget.com/whatis/feature/5-business-use-cases-for-NFTs.

NFTs and Tangible Assets: Transforming Real Estate and Beyond

The application of NFTs goes well beyond the digital realm, infiltrating the traditional domains of real estate and physical assets. [518]NFTs serve as digital representations of ownership, effectively streamlining and securing the transfer of tangible items.

NFT Titles: A token that cannot be copied or substituted, that is recorded on the blockchain that is used to certify the authenticity and ownership of a title bearing piece of real property or personal property.

In real estate, for instance, NFTs can represent property titles, providing a secure and transparent way to transfer ownership. The traditional, often cumbersome, process of title transfers can be revolutionized through the efficiency and security offered by blockchain technology and NFTs. This has the potential to simplify real estate transactions, reduce fraud, and enhance the overall accessibility of property ownership.[519]

Beyond real estate, luxury items such as watches, art, or even vehicles can be equipped with NFTs, providing a digital title that encapsulates authenticity and ownership history. This not only adds value to the items but also ensures a transparent and traceable provenance.

[518] Petre, Tom (2023). "From Tokens to Titles: How NFTs are Transforming the Real Estate Industry." FAUN at https://faun.pub/from-tokens-to-titles-how-nfts-are-transforming-the-real-estate-industry-dc7ac06f0d32.

[519] *Id.*

The Tokenization of Representation: NFTs as Digital Certificates

NFTs, in their essence, function as digital certificates of ownership and representation. Each token carries a unique identifier, often linked to metadata describing the item it represents.[520] This metadata can include details such as the item's origin, characteristics, ownership history, and more.

In the art world, NFTs have become synonymous with digital certificates of authenticity. Artists can tokenize their creations, turning them into unique NFTs that prove ownership and authenticity. This not only empowers artists but also opens up new possibilities for buyers and collectors to confidently engage in the digital art market.

While the integration of NFTs for titles and item representation brings about revolutionary possibilities, it is not without challenges. Regulatory frameworks, especially in industries like real estate, may need to adapt to accommodate blockchain-based ownership.[521] Ensuring legal recognition and protection of NFT-based titles is a critical consideration.

Interoperability and standards for representing items through NFTs also require attention. Establishing widely accepted protocols can enhance the fluidity of asset representation and transfer across different platforms and industries.[522]

---

[520] Gygi, Mattea (2022). "TreuMint helps mint the first commercial real estate title as an NFT." Tech Buzz at https://techbuzz.news/trumint-helps-mint-the-first-commercial-real-estate-title-as-an-nft/.
[521] *Id.*
[522] Bernstein, *supra*.

Conclusion: NFTs as the Vanguard of Ownership and Representation

In conclusion, NFTs stand as the vanguard of a new era in ownership and representation. Whether in the digital realm of gaming, the traditional landscape of real estate, or the world of luxury assets, NFTs are redefining how titles are conceived and how items are represented. The transformative power of blockchain technology, coupled with the uniqueness and security of NFTs, is poised to reshape industries and empower individuals in unprecedented ways. As we navigate this intersection of technology, ownership, and representation, the full potential of NFTs in titles and item representation is yet to be fully realized, promising an exciting journey into the future of ownership documentation.

Chapter 99: NFTs as Memorials

Introduction

Digital Artifacts: Digital representations of physical world events and

In the frontier of digital innovation, Non-Fungible Tokens (NFTs) have emerged as more than just cryptographic assets; they have become the vanguards of immortalizing moments and events. This chapter explores the profound impact of NFTs on the concept of event memorabilia, transcending traditional forms of documentation to create a digital legacy that is secure, verifiable, and everlasting.

Traditionally, event memorabilia has taken the form of physical items such as tickets, posters, and tangible souvenirs. These artifacts capture the essence of a moment in time, serving as mementos for those who participated in or witnessed significant events. While these physical tokens hold sentimental value, they are susceptible to wear, loss, and deterioration over time.

NFTs as Digital Artifacts: Redefining Memorabilia in the Digital Age

Enter NFTs, the digital counterparts that revolutionize the concept of event memorabilia. NFTs provide a unique and secure way to encapsulate moments, turning them into digital artifacts that can be owned, traded, and preserved on the blockchain. These digital assets carry the torch of tradition while introducing a new era of accessibility, authenticity, and permanence.

Event photography, a staple in capturing the vibrancy of moments, finds a new dimension with NFTs. Photographs from significant events can be tokenized, transforming them into limited edition NFTs that represent a slice of history.[523] Owners of these NFTs not only possess a digital copy of the image but also hold a stake in the event's legacy.

NFTs can imbue event photos with additional layers of information through metadata, including details about the event, the photographer, and even the individuals captured in the image. This metadata becomes an integral part of the NFT, enriching the overall storytelling aspect of the digital artifact.

Documenting Events on the Blockchain: Certainty in an Uncertain World

The blockchain, acting as an immutable ledger, provides the ideal canvas for documenting events through NFTs. The transparency and security inherent in blockchain technology ensure that the information and ownership details associated with event NFTs remain tamper-proof. This introduces a level of certainty and trust in the provenance of digital artifacts, assuring collectors and enthusiasts of the authenticity of their memorabilia.

Beyond photographs, artists can contribute to the realm of event memorabilia by creating unique renditions or interpretations of significant moments. These artistic expressions can be tokenized as NFTs, offering a fusion of creativity and historical significance. From digital paintings to 3D representations, artists can leverage NFTs to contribute to the ecosystem of event memorabilia in innovative ways.

Collectors, drawn by a passion for preserving history and a love for specific events, can now acquire NFTs as a form of event

---

[523] Staff (2023). "Memorial NFTs: A New Age Tribute to our Departed Loved Ones." FatMan at https://medium.com/@wordbaby.buzz/memorial-nfts-a-new-age-tribute-to-our-departed-loved-ones-7f8c3ec31c0f.

memorabilia. The limited nature of these digital artifacts adds a layer of exclusivity, turning ownership into a meaningful and often communal experience. NFT collectors become custodians of moments, creating a digital legacy that can be shared, traded, or passed down through generations.

Conclusion:

In conclusion, NFTs have transcended their role as mere digital assets; they have become time capsules that encapsulate the essence of human experience during significant events. Whether through photographs, artistic renditions, or unique interpretations, NFTs redefine how we document, share, and immortalize moments in the digital age. The intersection of technology, art, and cultural preservation converges in the world of event NFTs, forging a path towards a future where the memories we hold dear are secured, accessible, and everlasting. As we navigate this transformative landscape, the marriage of NFTs and event memorabilia paves the way for a new era of storytelling, where every token represents not just a moment, but a piece of the collective human narrative.

Chapter 100: NFTs as Smart Contracts.

Introduction

The convergence of Non-Fungible Tokens (NFTs) and smart contracts marks a transformative moment in the landscape of decentralized technologies. This chapter delves into the profound implications of using NFTs as smart contracts, exploring the synergies that redefine how we conceptualize, create, and execute digital agreements in the blockchain era.[524]

Smart contracts, programmable self-executing contracts with the terms of the agreement directly written into code, have long been a cornerstone of blockchain technology. These contracts automate and enforce the execution of predefined rules, eliminating the need for intermediaries in various transactions.[525] As smart contracts evolved beyond their utilitarian origins, their integration with NFTs introduced a new dimension, transforming code into collectibles.

NFTs as Programmable Assets: A New Layer of Functionality

While traditional NFTs have primarily represented static digital or digitized physical assets, the marriage of NFTs and smart contracts introduces programmability.[526] NFTs can now embody functionality, morphing from static tokens into interactive assets governed by the logic embedded in smart contracts. This programmability opens the door to a myriad of possibilities,

---

[524] Staff (2023A). "What is an NFT Smart Contract?" Hedra at https://hedera.com/learning/smart-contracts/nft-smart-contract.
[525] Delmegani, Cem (2024). "A Deep Dive into Smart Contract NFTs in 2024." AI Multiple at https://research.aimultiple.com/smart-contract-nft/.
[526] *Id.*

shaping the nature of ownership, royalties, and interactivity associated with digital assets.[527]

Ownership and Access Control: Smart Contracts Redefining Possession

Smart contracts embedded within NFTs redefine ownership s. Traditional ownership often implies a binary possession—either you own the asset or you don't.[528] With NFTs as smart contracts, ownership becomes multifaceted. Smart contracts can dictate specific access controls, allowing owners to grant or restrict certain privileges associated with the NFT.[529] This granular control over ownership introduces a nuanced layer of interaction and engagement with digital assets.

One of the revolutionary aspects of NFTs as smart contracts is the automated royalty system. Smart contracts can embed royalty clauses, ensuring that creators receive a percentage of the proceeds each time the NFT changes hands in the secondary market. This self-executing royalty mechanism addresses long-standing concerns about artists' compensation in the digital realm, creating a sustainable revenue stream tied directly to the success of their work.[530]

---

[527] Staff (2023). "Smart Contracts for NFTs." RV at https://www.realvision.com/blog/smart-contracts-for-nfts.

[528] Andrew, Gretchen (2022). "NFTs as 'smart contracts'- but what exactly are they?" The art newspaper at https://www.theartnewspaper.com/2022/08/17/nfts-use-smart-contractsbut-what-exactly-are-they.

[529] Hedra (2023A), *supra*.

[530] Ander, *supra*.

Conditional Logic: NFTs Responding to External Triggers

Smart contracts empower NFTs with conditional logic, enabling them to respond ally to external triggers or events.[531] For instance, an NFT representing a concert ticket could be programmed to unlock exclusive content or experiences based on specific conditions—attendance at the event or reaching a certain milestone. This interactivity transforms NFTs into living assets, evolving based on predefined rules and real-world occurrences.

The immutability of blockchain ensures that smart contracts, once deployed, cannot be altered retroactively.[532] This characteristic introduces a high level of trust and transparency into digital agreements. When NFTs embody smart contracts, the terms and conditions associated with the asset are not only self-executing but also tamper-proof, providing all parties involved with a secure and unalterable record of the agreement.

While the fusion of NFTs and smart contracts holds immense promise, it also presents challenges. The complexity of creating and understanding smart contracts can be a barrier for mainstream adoption.[533] Striking a balance between robust functionality and user-friendly interfaces is crucial to ensuring that the benefits of programmable NFTs are accessible to a broader audience.

Conclusion

In conclusion, the integration of NFTs and smart contracts ushers in a new era of programmable digital assets, redefining the landscape of ownership, royalties, and interactivity in the

---

[531] Delmegani, *supra*.
[532] Ray, Shaan (2021). "NFTs as Smart Contracts." Lansaar at https://medium.com/lansaar/nfts-and-smart-contracts-6c4c5516d5a0.
[533] *Id.*.

blockchain space. The symbiotic relationship between NFTs and smart contracts creates a paradigm shift in how we conceive and execute digital agreements. As we navigate this transformative landscape, the fusion of these decentralized technologies not only enriches the functionalities of digital assets but also paves the way for innovative applications across various industries. The journey of NFTs as smart contracts is an unfolding narrative, where the programmable nature of these assets propels us into a future where the code is not just law but a and responsive force shaping the digital economy.

Conclusion Part XI

Part XI of our exploration has propelled us into the cutting-edge realms of blockchain technology, specifically focusing on the transformative concepts of Non-Fungible Tokens (NFTs) and smart contracts. As we conclude this segment, it is evident that we have traversed the frontiers of digital innovation, where the convergence of decentralized technologies is reshaping paradigms and redefining possibilities.

Chapters 96 to 100 have meticulously unraveled the multifaceted canvas of NFTs, shedding light on their diverse applications and functionalities. From the conceptual introduction of NFTs to their nuanced roles as digital receipts, representations of items, memorials of events, and finally, as smart contracts—each chapter has been a brushstroke painting a vivid picture of the transformative power embedded in these unique tokens.

The synergy between smart contracts and NFTs emerged as a central theme, weaving together the autonomy of programmable agreements with the uniqueness and ownership representation encapsulated in NFTs. The integration of these two pillars of blockchain technology has not only introduced a paradigm shift in how we conceive digital assets but has also addressed longstanding challenges such as artist compensation, ownership s, and conditional interactivity.

At the intersection of trust and innovation lies the heart of blockchain's promise. Smart contracts, with their self-executing logic and immutability, bring an unparalleled level of transparency and reliability to digital agreements. NFTs, as the vessels for these agreements, add a layer of uniqueness and ownership representation that was previously elusive in the digital realm.

While our exploration delved into the immense potential of NFTs and smart contracts, it is imperative to acknowledge the challenges and considerations that accompany these advancements. Balancing complexity with accessibility, ensuring inclusivity in adoption, and addressing environmental concerns are crucial facets that demand ongoing attention as we navigate the ever-evolving landscape of decentralized technologies.

As we reflect on the insights gained from Chapters 96 to 100, it becomes clear that we are witnessing the dawn of a new era in digital innovation. The vision for the future extends beyond programmable assets, encompassing a spectrum of possibilities where blockchain technology permeates industries, reshapes economies, and empowers individuals on a global scale.

In conclusion, Part XI has been a journey through the avant-garde of blockchain technology, where the fusion of NFTs and smart contracts has unlocked doors to unprecedented opportunities. As we stand on the threshold of what lies ahead, the narrative of decentralized technologies continues to unfold, promising a future where innovation is not only driven by code but shaped by the interactions of programmable assets and the collective imagination of a connected world.

Part XII: SFTs and the Gap Between Traditional Tokens and NFTs

Introduction Part XII

In the ever-evolving landscape of blockchain technology, Chapters 101-105 embark on a comprehensive exploration into the uncharted territories of Semi-Fungible Tokens (SFTs). As the cryptoverse continues its transformative journey, these chapters unravel the intricacies, potentials, and applications of SFTs, propelling us into a realm where fungibility converges with traceability, and the traditional paradigms of finance and ownership undergo a paradigm shift.

Our journey begins with the fundamental demystification of Semi-Fungible Tokens (SFTs) in Chapter 101. Breaking down the technical nuances and conceptual underpinnings, we delve into what makes these tokens semi-fungible, exploring the unique attributes that set them apart in the vast spectrum of tokenization. The aim is to lay a solid foundation, ensuring a nuanced understanding as we navigate through the subsequent chapters. Chapter 102 casts a spotlight on the fungibility perspectives of SFTs. Here, we dissect how these tokens, with their semi-fungible nature, challenge and reshape traditional notions of fungibility within the cryptoverse. As we unravel the intricacies of fungibility, we gain insights into how SFTs carve a distinctive niche in the diverse landscape of digital assets.

Chapters 103 and 104 propel us into the heart of financial systems, exploring how SFTs can seamlessly integrate into traditional frameworks. From securities management to the revolutionizing of derivative ownership systems, our exploration extends beyond the boundaries of blockchain, envisioning a future where SFTs play a pivotal role in reshaping how we perceive and interact with financial instruments. Chapter 105 elevates the discourse, steering our focus towards the fusion of SFTs with traceable tokenization. Here, we explore the multifaceted applications of traceability, spanning supply chains, healthcare, finance, and intellectual property. The chapter not only reveals how SFTs contribute to transparency and accountability but also highlights the transformative potential of traceable tokenization in various sectors.

Throughout these chapters, a thematic interplay emerges – the delicate equilibrium between fungibility and traceability. As we navigate the unexplored frontiers of SFTs, it becomes apparent that these tokens represent more than just a technological innovation. They embody a harmonious balance, where the interchangeable nature of tokens coexists with the transparency and accountability introduced by traceability. Anticipation of challenges is inherent to progress. Chapters 101-105 pragmatically address considerations around privacy, scalability, regulatory compliance, and interoperability. By acknowledging these challenges, we adopt a forward-looking perspective, ensuring that the development and implementation of SFTs occur with a keen awareness of potential hurdles.

As we embark on this journey through Chapters 101-105, the overarching vision extends beyond individual chapters. It envisions collaborative futures where technological innovation, regulatory frameworks, and industry collaborations converge to shape the trajectory of SFTs. This collaborative spirit becomes instrumental in realizing the full potential of SFTs in a digital landscape characterized by decentralization, transparency, and accountability.

In essence, these chapters serve as a gateway to a continuum of discovery and refinement, where the horizons of tokenization extend further, unveiling new possibilities and applications for SFTs. The narrative woven through Chapters 101-105 invites readers to navigate this uncharted territory, fostering a deeper understanding of SFTs and their transformative impact on blockchain technology and the broader digital ecosystem.

Chapter 101: Introduction to SFTs

Introduction: The Intricacies of Semi-Fungibility

In the ever-evolving landscape of blockchain technology, Semi-Fungible Tokens (SFTs) emerge as a distinctive class of digital assets, offering a nuanced perspective on fungibility.[534] This chapter embarks on a journey to unravel the essence of SFTs, providing a basic yet insightful explanation of their characteristics and functionalities.

Before delving into the specifics of SFTs, it is crucial to grasp the concept of fungibility itself. Fungibility refers to the interchangeable nature of tokens within a particular class, where each unit is indistinguishable and holds an equal value.[535] Traditionally, this concept is associated with assets like fiat currencies or basic cryptocurrencies, where each unit is identical and interchangeable.

Semi-fungibility introduces a nuanced layer to the concept of interchangeability. Unlike fully fungible tokens, where each unit is identical, SFTs operate on a spectrum. This chapter explores how SFTs retain certain fungible qualities while allowing for unique distinctions between individual tokens. The semi-fungible nature of these tokens opens up possibilities for a more tailored and flexible digital asset framework.[536]

The Programmable s of SFTs

At the core of SFTs lies the programmable nature afforded by smart contracts. These contracts enable developers to embed

---

[534] Wells, Evan (2023). "An introduction to Semi-Fungible Tokens." Chain safe at https://blog.chainsafe.io/an-introduction-to-semi-fungible-tokens/.

[535] Antonovici, *supra.*

[536] Werkheiser, Brady (2023). "What are semi-fungible tokens." Alchemy at https://www.alchemy.com/overviews/semi-fungible-tokens-sfts.

specific functionalities into each token, adding a layer of uniqueness while maintaining a degree of interchangeability.[537] This section delves into the programmable s that contribute to the semi-fungible essence of these tokens, allowing for diverse use cases within blockchain ecosystems.

SFTs find applications across various domains, ranging from digital collectibles and gaming assets to real-world representations like tokenized real estate. By exploring the diverse use cases, this chapter sheds light on how the semi-fungible nature of these tokens caters to specific scenarios where individuality matters while retaining the benefits of fungibility.[538]

Interoperability, the seamless interaction between different blockchain networks, plays a crucial role in shaping the semi-fungible landscape. As SFTs traverse diverse platforms, their semi-fungible nature relies on effective standards for interoperability.[539] This section navigates through the relationship between interoperability and the semi-fungibility of tokens, emphasizing the importance of a cohesive blockchain ecosystem.

Conclusion: Embracing the Versatility of SFTs

In conclusion, it can be a good idea for business to embrace the versatility of Semi-Fungible Tokens. Positioned at the intersection of uniqueness and interchangeability, SFTs offer a framework for digital assets. Their semi-fungible nature opens avenues for creativity and customization within blockchain ecosystems, fostering a new wave of possibilities for developers,

---

[537] Wells, *supra*.

[538] Antonovici, *supra*.

[539] Dimegani, Cem (2024). "Semi-Fungible Tokens in 2024: Benefits and Use Cases." AI Multiple at https://research.aimultiple.com/semi-fungible-tokens/.

businesses, and users alike. As we traverse the landscape of blockchain innovation, SFTs stand as a testament to the ongoing evolution of tokenomics and decentralized systems.

Chapter 102: Fungibility Perspectives in SFTs

Introduction: Unveiling the Fungibility Spectrum

In the realm of the cryptoverse, where digital assets and blockchain technology converge, the concept of fungibility becomes a key issue. This chapter embarks on a journey to dissect the s of fungibility within the cryptoverse and sheds light on how Semi-Fungible Tokens (SFTs) represent a paradigm shift in this foundational aspect of tokenomics.

Fungibility, the property of interchangeability between units of a particular asset, has been a fundamental pillar in the cryptoverse. Traditional cryptocurrencies like Bitcoin and Ethereum operate on the principles of full fungibility, where each unit is identical, ensuring seamless transactions and market liquidity.[540] This section provides an overview of fungibility's historical significance and its role in shaping the functionality of digital assets.

The Fungibility Spectrum Unveiled

While traditional cryptocurrencies adhere to the principles of full fungibility, the cryptoverse has witnessed the emergence of a more nuanced spectrum. Semi-Fungible Tokens (SFTs) introduce a paradigm shift by offering a middle ground—retaining some fungible characteristics while allowing for unique distinctions between tokens.[541] This chapter delves into the implications of this spectrum, exploring the novel opportunities and challenges it brings to the forefront.[542]

SFTs represent a significant departure from the all-or-nothing fungibility model.[543] This section dissects how SFTs usher in a

---

[540] *Id.*

[541] Werkeiser, *supra.*

[542] Antonovici, *supra.*

[543] Staff (2022). "A Quick Guide to understanding Semi-Fungible Tokens." Oodles Blockchain at

paradigm shift by providing a more flexible and customizable approach to digital assets. The programmable nature of SFTs, embedded within smart contracts, allows developers to imbue individual tokens with distinct features, opening new possibilities for innovation and token use cases.

To comprehend the paradigm shift introduced by SFTs, it's essential to explore their diverse use cases within the cryptoverse.[544] From digital collectibles and gaming assets to tokenized real-world representations, SFTs cater to a broad spectrum of applications. This section navigates through these varied use cases, illustrating how the semi-fungible nature of tokens can be leveraged for specific scenarios where uniqueness and interchangeability coexist.

Challenges and Opportunities in Semi-Fungibility

As with any paradigm shift, the introduction of semi-fungibility brings forth both challenges and opportunities. From standardizing interoperability to ensuring a balanced ecosystem, this chapter discusses the s surrounding SFTs. By examining the potential hurdles and avenues for growth, it aims to provide a comprehensive understanding of how semi-fungibility is reshaping the landscape of digital assets.

Conclusion: Redefining Fungibility in the Cryptoverse

In conclusion, this chapter discusses the differences between fungibility and uniqueness within the cryptoverse, showcasing how Semi-Fungible Tokens (SFTs) redefine this fundamental aspect of digital assets. As SFTs bridge the gap between full

---

https://blockchain.oodles.io/blog/understanding-semi-fungible-tokens/.

[544] Staff (2023). "What are Semi-Fungible Tokens, And How are they different from NFTs?" Protlitus Technologies at https://medium.com/@Prolitus01/what-are-semi-fungible-tokens-and-how-are-they-different-from-nfts-ca67a52d33c6.

fungibility and individual distinctiveness, they open a new chapter in the evolution of tokenomics. Navigating through the challenges and embracing the opportunities, SFTs exemplify the adaptability and innovation inherent in blockchain technology. The paradigm shift introduced by SFTs prompts us to rethink the conventional notions of fungibility, paving the way for a more versatile and   cryptoverse.

Chapter 103: SFTs and Stock Systems

Introduction: Transforming Securities with Semi-Fungibility

As the financial markets and securities management systems evolve, this chapter embarks on a journey to explore the potential integration of Semi-Fungible Tokens (SFTs). This chapter delves into the transformative possibilities that SFTs offer for enhancing the efficiency, transparency, and accessibility of stock systems in the realm of securities management.[545]

To comprehend the impact of SFTs on stock systems, it is crucial to understand the evolution of securities management.[546] Traditional stock systems have long operated on centralized models, introducing challenges related to intermediaries, settlement times, and accessibility. This section provides a concise overview of the existing stock systems and the inherent limitations that the integration of SFTs aims to address.

Unlocking Efficiency Through SFTs: The Potential Benefits

SFTs bring forth a myriad of potential benefits when integrated into stock systems for securities management. This segment elucidates how the semi-fungible nature of tokens can streamline processes such as issuance, transfer, and trading of securities.[547] Enhanced liquidity, reduced settlement times, and lower transaction costs are among the benefits that SFTs can bring to the traditional stock ecosystem.

---

[545] Antonovici, *supra*.

[546] Grupta, Ketan (2023). "SFT- Semi-Fungible Tokens: Implementation, Benefits & Use Cases." Medium at https://kettan007.medium.com/sft-semi-fungible-tokens-implementation-benefits-use-cases-45801388dcc9.

[547] *Id.*

Amidst the promising benefits, it is paramount to address the regulatory landscape surrounding securities management. This chapter navigates the complexities of compliance and regulatory alignment when incorporating SFTs into stock systems.[548] By examining how smart contracts and blockchain technology can facilitate regulatory adherence, it sheds light on the path towards a harmonious integration.

SFTs have the potential to democratize access to securities by lowering entry barriers and fostering inclusivity.[549] This section explores how fractional ownership, enabled by the semi-fungible nature of tokens, can empower a broader spectrum of investors. From retail participants to institutions, the integration of SFTs aims to create a more accessible and inclusive securities market.

Challenges on the Horizon: Navigating the Transition

As with any transformative technology, the integration of SFTs into stock systems comes with its set of challenges.[550] From technological hurdles to market acceptance, this chapter examines the obstacles that need to be navigated during the transition.[551] Addressing concerns related to interoperability, standardization, and investor education, it provides a comprehensive view of the road ahead.

Conclusion: Paving the Way for Future Securities Management

In conclusion, this chapter illuminates the transformative potential of integrating Semi-Fungible Tokens into stock systems

---

[548] Antonovici, *supra*.
[549] *Id*.
[550] Staff (2023). "Semi-Fungible Token Development Platforms." Prolitus at https://www.prolitus.com/semi-fungible-token-development/.
[551] *Id*.

for securities management. As the financial landscape undergoes a paradigm shift, SFTs emerge as a catalyst for efficiency, inclusivity, and transparency. By unlocking the benefits of blockchain technology and smart contracts, SFTs pave the way for a future where securities are more accessible, liquid, and seamlessly traded. Navigating challenges and building on the successes of early adopters, the integration of SFTs represents a pivotal moment in the evolution of securities management, promising a more and democratized future.

Chapter 104: SFTs and Derivative Ownership

Introduction

In the landscape of financial instruments, this chapter embarks on a comprehensive exploration of how Semi-Fungible Tokens (SFTs) have the potential to revolutionize derivative ownership systems. Derivatives, long confined to traditional frameworks, stand at the precipice of transformation through the unique features offered by SFTs.[552] This chapter delves into the fundamental shifts, benefits, and challenges that accompany the integration of SFTs in the realm of derivative ownership.

Before delving into the transformative potential of SFTs, it is imperative to establish a foundational understanding of derivative ownership systems.[553] This section provides an overview of derivatives, their types, and the traditional mechanisms governing derivative ownership. By examining the existing challenges and limitations, it sets the stage for the disruptive impact that SFTs can introduce.

SFTs bring forth a set of unique attributes that can act as catalysts for fundamental change in derivative ownership systems. This segment elucidates how the semi-fungible nature of tokens can redefine the issuance, trading, and management of derivatives.[554] Enhanced liquidity, real-time settlement, and fractional ownership are among the transformative features that SFTs introduce to the derivative landscape.

---

552 Staff (2023). "Semi-Fungible Token Development- (SFTs) The Future of Crypto World." Coinsclone https://www.coinsclone.com/semi-fungible-token-development/.

553 Gate.io (2023). "What are Semi-Fungible Tokens?" at https://www.gate.io/blog_detail/1199/what-are-semi-fungible-tokens.

554 *Id.*

Tokenizing Derivatives: Benefits and Opportunities

Tokenization of derivatives through SFTs opens up a plethora of benefits and opportunities. This section explores how the tokenization process enhances accessibility, reduces counterparty risk, and introduces programmable functionalities to derivatives. By dissecting specific use cases and examples, it paints a vivid picture of the advantages that SFTs bring to derivative ownership systems.

At the heart of the SFT revolution lies the integration of smart contracts and automation into derivative ownership systems.[555] This chapter delves into the transformative potential of programmable contracts in derivatives, automating processes such as execution, settlement, and risk management. By examining real-world implementations, it showcases the efficiency gains and risk mitigation achieved through smart contract integration.

While the benefits are compelling, the transition to SFT-driven derivative ownership systems is not without challenges. This segment navigates through potential hurdles such as regulatory considerations, interoperability issues, and market acceptance.[556] By providing strategies to address these challenges, it offers a roadmap for market participants and regulators during the transformative journey.

SFTs inherently embed security features through blockchain technology, ensuring transparency and trust in derivative ownership. This section explores how the immutability of blockchain ledgers and cryptographic security measures

---

[555] Williams, Catherine (2023). "Semi-Fungible Tokens: The New Standard for Tokenized Assets." CoinMonks at https://medium.com/coinmonks/semi-fungible-tokens-the-new-standard-for-tokenized-assets-379fb93247f9.

[556] *Id.*

contribute to a more secure and transparent derivative ecosystem. It also examines the role of decentralized networks in mitigating risks associated with centralized clearinghouses.

Conclusion:

In conclusion, this chapter unravels the transformative potential of Semi-Fungible Tokens in revolutionizing derivative ownership systems. As the financial landscape undergoes a paradigm shift, SFTs emerge as a pioneering force, reshaping how derivatives are issued, traded, and managed. By harnessing the power of blockchain, smart contracts, and automation, SFTs introduce unparalleled efficiency, transparency, and security to the derivative ecosystem. While challenges exist, the benefits outweigh them, paving the way for a future where derivatives are more accessible, liquid, and seamlessly integrated into the digital financial infrastructure. The integration of SFTs in derivative ownership marks a pivotal moment in financial innovation, promising a and decentralized future for derivative markets.

Chapter 105: Traceable Tokenization with SFTs

Introduction: Navigating the Evolution of Tokenization with Traceability

In the ever-evolving landscape of blockchain technology and tokenization, Chapter 105 embarks on a comprehensive exploration of how traceable tokenization, facilitated by Semi-Fungible Tokens (SFTs), is shaping a new paradigm.[557] This chapter delves into the concept of traceability within tokenization, unraveling the layers of transparency, accountability, and security that traceable SFTs bring to various domains.

To comprehend the significance of traceable tokenization, it is crucial to establish a foundational understanding of the term.[558] This section provides insights into the core principles of traceability within tokenization and outlines the key objectives it aims to achieve. By examining historical precedents and use cases, it sets the stage for a nuanced exploration of traceable SFTs.

Semi-Fungible Tokens (SFTs) emerge as key enablers in achieving traceability within tokenized ecosystems. This segment explores the unique features of SFTs that contribute to traceability, such as enhanced metadata capabilities, cryptographic security, and programmable functionalities.[559] By dissecting the underlying mechanisms, it sheds light on how SFTs serve as the bulwark for traceable tokenization.

---

[557] Staff (2023). "What is 'Semi-Fungible' Crypto Token." Coingape at https://coingape.com/blog/what-is-semi-fungible-crypto-token/.
[558] *Id.*
[559] Oodles, *supra.*

Applications Across Industries: Traceability as a Cross-Cutting Imperative

The concept of traceable tokenization transcends industry boundaries, finding applications in diverse sectors. This section navigates through use cases across industries, including supply chain management, healthcare, finance, and intellectual property.[560] By showcasing how traceable SFTs address industry-specific challenges, it underscores the versatility and adaptability of traceability in diverse domains.

One of the primary domains where traceable SFTs make a profound impact is in supply chain management. This segment delves into how SFTs enable end-to-end traceability of products, from raw materials to the final consumer.[561] By enhancing transparency and accountability, traceable SFTs revolutionize supply chain processes, mitigating issues like counterfeiting, fraud, and inefficiencies.[562]

In the realm of healthcare, the traceability offered by SFTs extends to the secure management of patient records and sensitive medical data.[563] This section explores how traceable SFTs contribute to data integrity, interoperability, and patient privacy. It delves into the potential of decentralized health records to empower individuals while ensuring compliance with stringent data protection regulations.

---

[560] Ambolis, Diana (2024). "Top 7 uses of 'Semi-Fungible' Crypto tokens that make sense." Blockchain Magazine at https://blockchainmagazine.net/top-7-use-cases-of-semi-fungible-crypto-tokens-that-make-sense/.
[561] *Id.*
[562] Staff (2023). "Why Carbonable Opts for Semi-Fungible Tokens." Carbonable.io at https://carbonable.medium.com/why-carbonable-opts-for-semi-fungible-tokens-7014d0672964
[563] Id.

Financial Transactions and Auditing: Redefining Accountability

Traceable SFTs bring a paradigm shift to financial transactions and auditing processes. This segment examines how SFTs enhance transparency in financial transactions, enabling real-time auditing and reducing the risk of fraudulent activities.[564] By streamlining reconciliation and settlement procedures, traceable SFTs redefine accountability within the financial landscape.

In the digital age, intellectual property and digital assets demand robust mechanisms for traceability and provenance.[565] This section investigates how SFTs provide immutable records of ownership, origin, and transfers for digital creations. By safeguarding intellectual property rights and facilitating fair compensation, traceable SFTs emerge as a transformative force in the realm of digital assets.

Challenges and Considerations: Navigating the Path to Traceability

While traceable SFTs offer substantial benefits, they are not without challenges. This segment explores considerations related to privacy, scalability, regulatory compliance, and interoperability. By acknowledging these challenges, stakeholders can navigate the path to traceability more effectively, fostering responsible and sustainable implementations.

Conclusion

In conclusion, this chapter navigates the landscape of traceable tokenization powered by Semi-Fungible Tokens (SFTs). The evolution of blockchain technology, coupled with the unique

---

564 Williams, *supra.*

565 Coingape, *supra.*

features of SFTs, paves the way for a future where transparency, accountability, and traceability redefine digital ecosystems. Across industries, from supply chains to healthcare and finance, traceable SFTs emerge as a bulwark for secure and auditable transactions. While challenges exist, the potential benefits in mitigating fraud, ensuring data integrity, and redefining ownership paradigms position traceable SFTs as a transformative force in the broader narrative of blockchain adoption. As we chart the future of transparent and traceable tokenization, the collaborative efforts of technologists, regulators, and industry participants will play a pivotal role in realizing the full potential of SFTs in creating a more accountable and interconnected digital world.

Conclusion Part XII

In the world of blockchain evolution, Chapters 101-105 have ventured into the realms of tokenization, specifically focusing on Semi-Fungible Tokens (SFTs) and their diverse applications. This synthesis encapsulates the key takeaways and overarching themes that emerge from this exploration.

Chapters 101 and 102 laid the groundwork, demystifying the concept of Semi-Fungible Tokens (SFTs). From understanding their basic nature to delving into their fungibility perspectives, these chapters illuminated the transformative potential SFTs hold in reshaping the conventional notions of fungibility within the cryptoverse. Chapters 103 and 104 expanded the narrative by investigating how SFTs could seamlessly integrate into traditional financial systems. By examining their potential in securities management and revolutionizing derivative ownership systems, these chapters underscored the adaptability of SFTs in addressing complexities within established financial frameworks. Chapter 105 elevated the discourse, steering the focus towards traceable tokenization with SFTs. The exploration of traceability across various industries – supply chain, healthcare, finance, and intellectual property – demonstrated how SFTs serve as catalysts for transparency, accountability, and security in an interconnected world.

The interplay between fungibility and traceability emerged as a recurring theme. While fungibility preserves the interchangeable nature of tokens, traceability, as explored in Chapter 105, introduces a layer of transparency and accountability. The delicate equilibrium between these attributes positions SFTs as versatile instruments capable of navigating diverse use cases. Acknowledging challenges is a vital facet of progress. Throughout these chapters, considerations around privacy, scalability, regulatory compliance, and interoperability were highlighted. Recognizing these challenges becomes instrumental in steering the development and implementation of SFTs with a pragmatic and informed approach.

As we conclude this journey through Chapters 101-105, the vision of the future involves collaborative endeavors. The fusion of technological innovation, regulatory frameworks, and industry collaborations will play a pivotal role in realizing the full potential of SFTs. It's a future where decentralized, transparent, and accountable systems redefine the way we perceive and interact with digital assets. The exploration of SFTs and their applications is an ongoing journey. Beyond these chapters lies a continuum of discovery, experimentation, and refinement. The nature of blockchain and tokenization ensures that the horizons will continually expand, unveiling new possibilities and applications for SFTs.

Chapters 101-105 serve as waypoints in a broader narrative – a narrative that unfolds at the intersection of technology, finance, governance, and societal transformations. The promise of SFTs lies not just in their fungibility or traceability but in their potential to catalyze a paradigm shift towards a more inclusive, transparent, and efficient digital ecosystem. As we navigate these horizons, the collaborative efforts of visionaries, technologists, regulators, and industry leaders will shape the trajectory of SFTs, contributing to the evolution of blockchain technology and tokenization in the years to come.

Conclusion

The closing chapters of this expansive exploration converge upon a pivotal question: where is the future built, and what role do Blockchain, DeFi, Web3, Crypto, and the Internet of Things (IoT) play in shaping this digital frontier? As we navigate this concluding chapter, we embark on a journey that transcends individual technologies, envisioning a landscape where these elements harmonize to redefine how we interact, transact, and envision the very fabric of our digital existence.

1. The Blockchain Foundation: A Pillar of Trust and Decentralization

At the heart of our future narrative lies Blockchain, a foundational pillar that has redefined trust, security, and decentralization. With its immutable ledger and cryptographic consensus mechanisms, Blockchain ensures transparency, immutability, and trust in a trustless environment. The future built on Blockchain extends beyond cryptocurrency transactions, encompassing supply chain management, healthcare, voting systems, and beyond. As industries adopt blockchain solutions, the decentralized ethos becomes integral to the fabric of our digital interactions.

2. DeFi's Ascendancy: Redefining Financial Landscapes

Decentralized Finance (DeFi) emerges as a force reshaping traditional financial landscapes. In the future we envision, DeFi transcends its current role, expanding into mainstream finance. Automated smart contracts facilitate lending, borrowing, and trading without intermediaries, democratizing financial services. The concept of programmable money gains prominence, offering a glimpse into a financial ecosystem where traditional barriers dissolve, and financial inclusion becomes a global reality.

3. Web3: Bridging the Gap Between Users and the Internet

Web3 emerges as the natural evolution of the internet, emphasizing user sovereignty, data ownership, and decentralized applications (DApps). In the future built on Web3, users seamlessly interact with applications and platforms, retaining control over their data and digital identities. The architecture of the internet evolves, transitioning from centralized servers to decentralized networks, fostering a more equitable and user-centric digital experience.

4. Crypto's Maturation: From Speculation to Everyday Utility

Cryptocurrencies, once the domain of speculative trading, mature into everyday utilities in our envisioned future. Stablecoins facilitate daily transactions, and major cryptocurrencies become widely accepted as means of payment. The volatility that characterized early crypto markets gives way to stability, and regulatory frameworks provide a secure environment for mass adoption. The narrative shifts from crypto as an investment class to crypto as an integral part of our digital economy.

5. IoT's Integration: A Networked World of Smart Devices

The Internet of Things (IoT) becomes a pervasive force, transforming our physical surroundings into interconnected networks of smart devices. In this future, IoT devices seamlessly communicate, facilitating data exchange and automation. Blockchain's tamper-proof nature ensures the integrity and security of IoT data, fostering a trusted environment for the exponential growth of interconnected devices. From smart homes to industrial applications, IoT becomes an indispensable component of our daily lives.

6. Synergies Unleashed: The Convergence of Technologies

What distinguishes this future is the synergistic convergence of Blockchain, DeFi, Web3, Crypto, and IoT. These technologies cease to exist in isolation, instead forming a seamless ecosystem where each element complements and enhances the capabilities of the others. Blockchain ensures the security of DeFi transactions, Web3 provides the interface for user-friendly interactions, Cryptocurrencies lubricate the wheels of financial transactions, and IoT creates a networked environment where data flows seamlessly.

7. Challenges and Resilience: Navigating the Path Forward

As we gaze into this future, it's crucial to acknowledge the challenges that accompany such transformative visions. Scalability concerns, regulatory hurdles, and the need for widespread education demand attention. However, the resilience of these technologies lies in their adaptability. Industry collaboration, regulatory clarity, and ongoing technological innovation pave the way for overcoming obstacles and ensuring the sustainable growth of this converged ecosystem.

8. Democratization of Innovation: Inclusion as a Guiding Principle

A future built on these technologies is inherently democratic, ensuring that innovation is not confined to privileged echelons. Blockchain, DeFi, Web3, Crypto, and IoT collectively contribute to the democratization of opportunities, whether in finance, information access, or technological advancements. Inclusion becomes a guiding principle, eroding barriers that once hindered progress.

9. Ethical Considerations: Crafting a Responsible Tomorrow

In crafting this future, ethical considerations take center stage. Privacy, security, and responsible use of technology become non-negotiable principles. The decentralized nature of this envisioned future ensures that power is distributed, preventing the concentration of influence in the hands of a few. Transparency, fairness, and ethical frameworks guide the development and deployment of technologies.

10. Call to Action: Nurturing the Seeds of Transformation

As we conclude this exploration, it serves as a call to action. Nurturing the seeds of transformation requires collective effort, industry collaboration, and individual empowerment. Whether as developers, policymakers, or everyday users, our roles extend beyond passive participation. The future is a collective creation, and each action today shapes the contours of tomorrow.

In the system of Blockchain, DeFi, Web3, Crypto, and IoT, the future emerges as a cohesive heterogeneous ecosystem, where the decentralized ethos, financial inclusion, user sovereignty, and technological innovation converge. This narrative, weaving together the threads of various technologies, envisions a future where the digital landscape reflects the values of transparency, trust, and empowerment. The chapters may close, but the journey into this transformative future is an ongoing exploration, inviting all stakeholders to actively shape the destiny of our decentralized and interconnected digital realm.

Made in United States
North Haven, CT
04 November 2026

21764780R00261